The ESSENTIALS of

REGISTERED TRADEMARK

ELECTRONICS II

Staff of Research and Education Association,
Dr. M. Fogiel, Director

This book is a continuation of "*THE ESSENTIALS OF ELECTRONICS I*" and begins with Chapter 10. It covers the usual course outline of Electronics II. Earlier/basic topics are covered in "*THE ESSENTIALS OF ELECTRONICS I*".

Research and Education Association
61 Ethel Road West
Piscataway, New Jersey 08854

THE ESSENTIALS ®
OF ELECTRONICS II

Printed in the United States of America

Library of Congress Catalog Card Number 87-61809

International Standard Book Number 0-87891-592-3

REVISED PRINTING, 1993

WHAT "THE ESSENTIALS" WILL DO FOR YOU

This book is a review and study guide. It is comprehensive and it is concise.

It helps in preparing for exams, in doing homework, and remains a handy reference source at all times.

It condenses the vast amount of detail characteristic of the subject matter and summarizes the **essentials** of the field.

It will thus save hours of study and preparation time.

The book provides quick access to the important facts, principles, theorems, concepts, and equations of the field.

Materials needed for exams, can be reviewed in summary form — eliminating the need to read and re-read many pages of textbook and class notes. The summaries will even tend to bring detail to mind that had been previously read or noted.

This "ESSENTIALS" book has been carefully prepared by educators and professionals and was subsequently reviewed by another group of editors to assure accuracy and maximum usefulness.

Dr. Max Fogiel
Program Director

CONTENTS

This book is a continuation of *"THE ESSENTIALS OF ELECTRONICS I"* and begins with Chapter 10. It covers the usual course outline of Electronics II. Earlier/basic topics are covered in *"THE ESSENTIALS OF ELECTRONICS I"*.

Chapter No. **Page No.**

10 OPERATIONAL AMPLIFIERS 92

10.1 The Basic Operational Amplifier 92
10.2 Offset Error Voltages and Currents 94
10.3 Measurement of Operational Amplifier Parameters 96

11 OPERATIONAL AMPLIFIER SYSTEMS 100

11.1 Basic Operational Amplifier Applications 100
11.1.1 Sign Changer, or Inverter 100
11.1.2 Scale Changer 100
11.1.3 Phase Shifter 101
11.1.4 Adder 101
11.1.5 Non-Inverting Adder 101
11.1.6 Voltage-To-Current Converter (Transconductive Amplifier) 102
11.1.7 Current-To-Voltage Converter 102
11.1.8 D.C. Voltage Follower 103
11.2 A.C.-Coupled Amplifier 103
11.3 Active Filters 105
11.3.1 Ideal Filters 105
11.3.2 Butterworth Filter 105

12 **FEEDBACK AND FREQUENCY COMPENSATION OF OP AMPS** 108

12.1 Basic Concepts of FEEDBACK 108
12.2 Frequency Response of a FEEDBACK Amplifier 111
12.3 Stabilizing Networks 114

13 **MULTIVIBRATORS** 117

13.1 Collector-Coupled Monostable Multivibrators 117
13.2 Emitter-Coupled Monostable Multivibrators 120
13.3 Collector-Coupled Astable Multivibrators 124
13.4 Emitter-Coupled Astable Multivibrators 125

14 **LOGIC GATES AND FAMILIES** 128

14.1 Logic Level Concepts 128
14.2 Basic Passive Logic 130
14.2.1 Resistor Logic (RL) 130
14.2.2 Diode Logic Circuits 131
14.2.3 Logic Circuit Current, Voltage, and Parameter Definitions and Notation 131
14.3 Basic Active Logic 132
14.3.1 Resistor-Transistor Logic (RTL) 132
14.3.2 Diode-Transistor Logic (DTL) 133
14.4 Advanced Active Logic Gates 134
14.4.1 Transistor-Transistor Logic (TTL) 134
14.4.2 Emitter-Couple Logic (ECL) 137

15 **BOOLEAN ALGEBRA** 138

15.1 Logic Functions 138
15.1.1 NOT Function 138
15.1.2 AND Function 138
15.1.3 OR Function 139
15.2 Boolean Algebra 139
15.3 The NAND and NOR Functions 142
15.4 Standard Forms for Logic Functions 143
15.5 The Karnaugh Map 144

16 REGISTERS, COUNTERS AND ARITHMETIC UNITS 148

16.1 Shift Registers 148
16.1.1 Serial-In Shift Registers 148
16.1.2 Parellel-In Shift Registers 149
16.1.3 Universal Shift Registers 150
16.2 Counters 151
16.2.1 The Ripple Counter 151
16.2.2 The Synchronous Counter 152
16.3 Arithmetic Circuits 153
16.3.1 Addition of Two Binary Digits, The Half Adder 153
16.3.2 The Full Adder 154
16.3.3 Parallel Addition 155
16.3.4 Look-Ahead-Carry Adders 155

17 OSCILLATORS 156

17.1 Harmonic Oscillators 156
17.1.1 The RC Phase Shift Oscillator 156
17.1.2 The Colpitts Oscillator 157
17.1.3 The Hartley Oscillator 158
17.1.4 The Clapp Oscillator 159
17.1.5 The Crystal Oscillator 159
17.1.6 Tunnel Diode Oscillators 160
17.2 Relaxation Oscillators 162

18 RADIO-FREQUENCY CIRCUITS 164

18.1 Non-Linear Circuits 164
18.2 Small-Signal RF Amplifiers 167
18.3 Tuned Circuits 168
18.4 Circuits Employing Bipolar Transistors 169
18.5 Analysis Using Admittance Parameters 171

19 FLIP-FLOPS 173

19.1 Types of Flip-Flops 173
19.1.1 The Basic Flip-Flop 173
19.1.2 R-S Flip-Flop 173
19.1.3 Synchronous R-S Flip-Flop (Clocked R-S Flip-Flop) 174

19.1.4 Preset and Clear 175
19.1.5 D-Type Flip Flop 175
19.1.6 J-K Flip-Flop 176
19.1.7 T-Type Flip-Flop 176
19.1.8 Master-Slave Flip-Flops 177
19.2 Flip-Flop Timing 177
19.3 Collector-Coupled Flip-Flops 178
19.4 Emitter-Coupled Flip-Flops 179
19.5 Switching Speed of a Flip-Flop 180
19.6 Regenerative Circuits 182

20 WAVESHAPING AND WAVEFORM GENERATORS 184

20.1 Common Waveforms 184
20.2 Linear Waveshaping Circuits 185
20.3 Sweep Generators 191

CHAPTER 10

OPERATIONAL AMPLIFIERS

10.1 THE BASIC OPERATIONAL AMPLIFIER

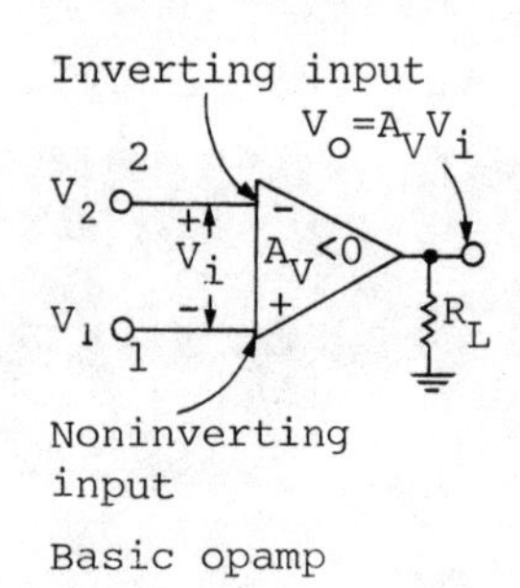

Basic opamp

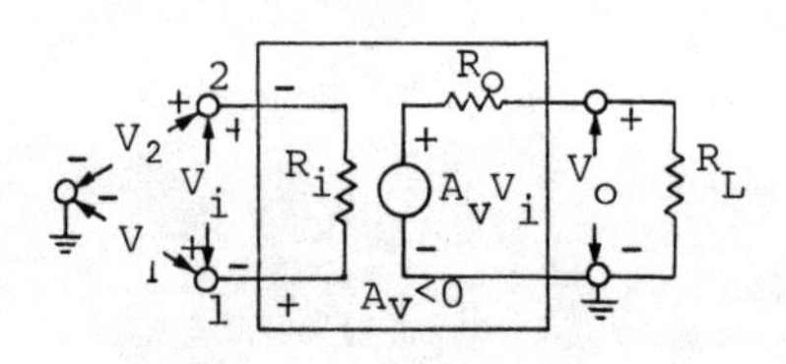

Low-frequency model of opamp

Ideal op amp:

A) $R_i = \infty$

B) $R_o = 0$

C) $A_v = -\infty$

D) B.W. = ∞

E) $V_o = 0$ when $V_1 = V_2$, independent of the magnitude of V_1

F) no drift of characteristics

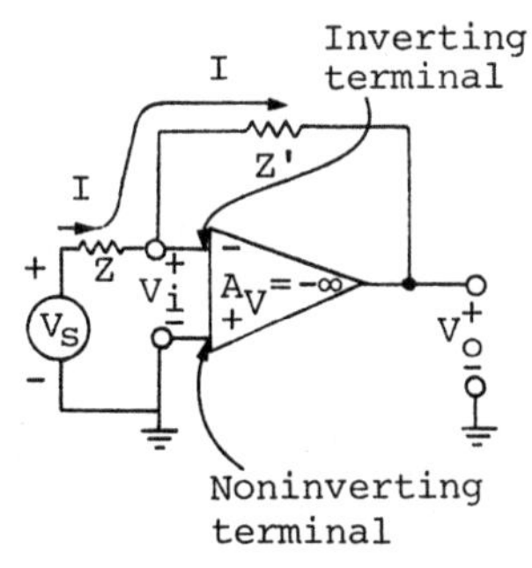

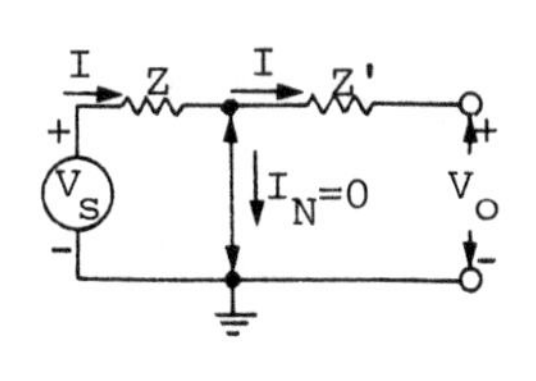

Ideal op amp with
feedback impedances

This is the basic inverting circuit. This topology represents voltage-shunt feedback.

$$A_{vf} = \text{voltage gain with feedback} = \frac{-Z'}{Z}$$

Inverting operational amplifier:

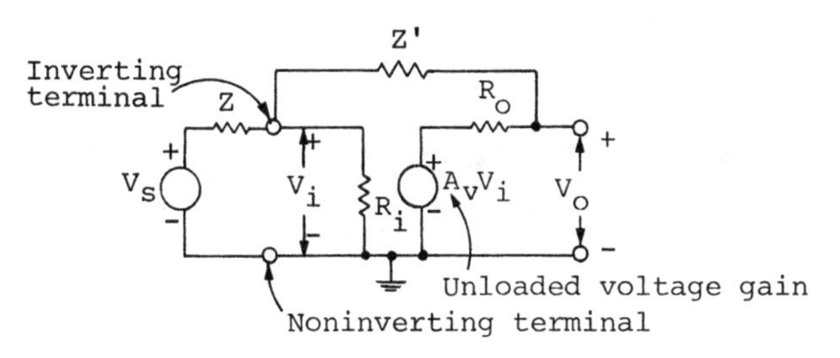

Small signal model

For a small-signal model, $|A_v| \neq \infty$, $R_i \neq \infty$ and $R_o \neq 0$.

$$A_{vf} = -y/y' - \left(\frac{1}{A_v}\right)(y'+y+y_i)$$

(where the y's are the admittances)

$$-A_v = \frac{V_o}{V_i} \text{ (with Z')} = \frac{A_v + R_o\ y'}{1 + R_o y'}$$

Non-inverting op amp:

The configuration is that of a voltage-series feedback amplifier, with the feedback voltage, v_f, equal to v_2.

The feedback factor $\beta = \frac{V_2}{V_o} = \frac{Z}{Z+Z'} \; (I_2 = 0)$

If $A_v \beta >> 1$, then

$$A_{vf} \approx 1/\beta = \frac{Z+Z'}{Z} = 1 + \frac{Z'}{Z}$$

Configuration:

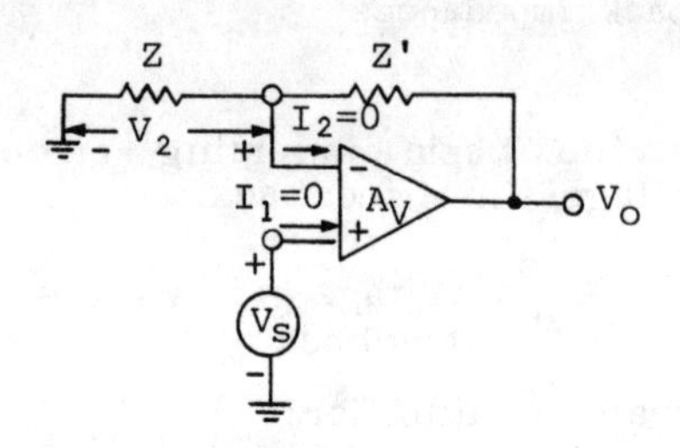

10.2 OFFSET ERROR VOLTAGES AND CURRENTS

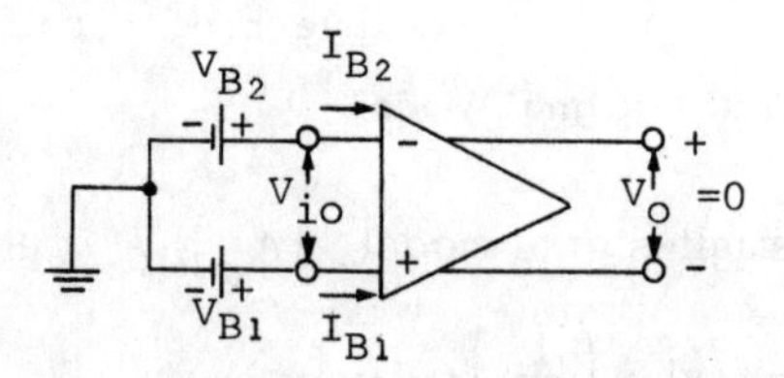

Input bias currents I_{B1} and I_{B2} and offset voltage V_{io}.

The input offset current is the difference between the input currents entering the input terminals of a balanced amplifier. In the figure above,

$$I_{io} = I_{B_1} - I_{B_2} \qquad \text{(when } V_o = 0\text{)}$$

Input offset current drift - This drift is described by the ratio $\frac{\Delta V_{io}}{\Delta T}$,where

ΔV_{io} = the change of input offset voltage

and ΔT = change in temperature.

Input offset voltage - When applied to the input terminals, this voltage will balance the amplifier.

Input offset voltage drift:

$$\frac{\Delta V_{io}}{\Delta T}$$

ΔV_{io} = change of input offset voltage.

Output offset voltage - This voltage marks the difference between the dc voltages measured at the output terminals on grounding the two input terminals.

Input common mode range - This is the range of the common mode input signal for which the differential amplifier remains linear.

Input differential range - This range is the maximum difference signal that can safely be applied to the op amp input terminals.

Output voltage range - This is the maximum output swing that can be obtained without significant distortion at a specified load resistance.

Full-power bandwidth - This bandwidth is the maximum frequency at which a sinusoid whose size is the output voltage range is obtained.

Slew rate - This is the time rate of change of the closed-loop amplifier output voltage under large signal conditions.

The model of an op amp and balancing techniques:

Model:

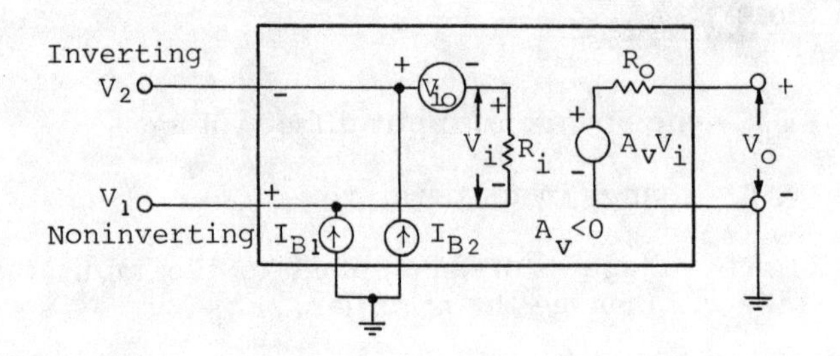

Universal balancing technique:

It is necessary to apply a small dc voltage in the input to bring the d output voltage to zero.

The following circuit supplies a small voltage in a series where, the non-inverting terminal is in the range $\pm V\left[\frac{R_2}{R_2+R_3}\right]$ = ±15mV, if ±15 V supplies are used.

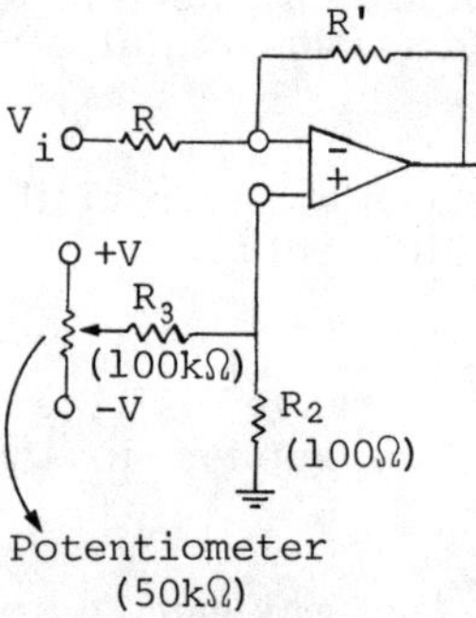

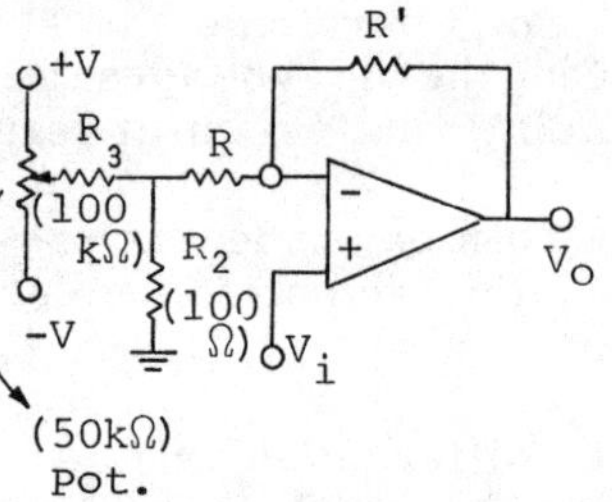

10.3 MEASUREMENT OF OPERATIONAL AMPLIFIER PARAMETERS

Input offset voltage V_{io} :

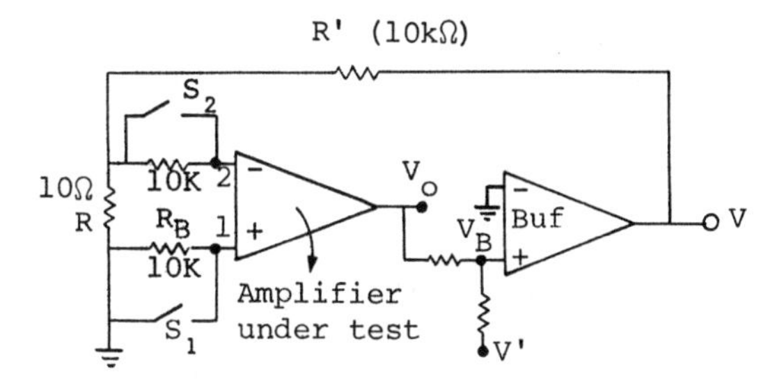

Set $V' = 0$ to get $V_o = 0$. The close s_1 and s_2.

If $V_o = 0$, then $V_i = 0$, and V_{io} appears between the inverting and non-inverting terminals.

$$V = \frac{V_{io}}{R}(R+R') = 1001\,V_{io} \approx 10^3 V_{io} \equiv V_3.$$

From the meter reading V_3 in volts, we get V_{io} in mv.

Power supply rejection ratio $= \dfrac{\Delta V_{io}}{\Delta V_{ce}}$ (ΔV_{io} and ΔV_{ce} the difference in the two input offset voltages)

Input bias current:

S_1 and S_2 are open and closed, respectively, and $V' = 0$.

Voltage across R $= V_{io} - R_B \cdot I_{B_1}$ and $V = \dfrac{R+R'}{R}(V_{io} - R_B \cdot I_{B_1})$

$$\approx 10^3 (V_{io} - 10^4 I_{B_1}) \equiv V_4$$

$-I_{B_1} = (V_4 - V_3)10^{-7}\text{A} = 100(V_4 - V_3)\text{mA}$

Open s_2 and close s_1; $V' = 0$ and we get I_{B_2}.

Bias current $I_B = \frac{1}{2}(I_{B_1} + I_{B_2})$ and I_{io}(offset) $= I_{B_1} - I_{B_2}$.

Open-loop differential voltage gain $A_v = A_d$:

S_1 and S_2 are closed, and V' is set to the output voltage = -10V then, V_o = -V' = 10v.

$$V = \frac{R+R'}{R}(V_{io}+V_i) \approx 10^3\left(V_{io} + \frac{V_o}{A_v}\right) \equiv V_5$$

$$A_v = \frac{10^3 \cdot V_o}{V_5 - V_3} = 10^4/V_5 - V_3$$

If we want to know the voltage gain A_v when there is a load, it is necessary to place R_L between V_o and the ground.

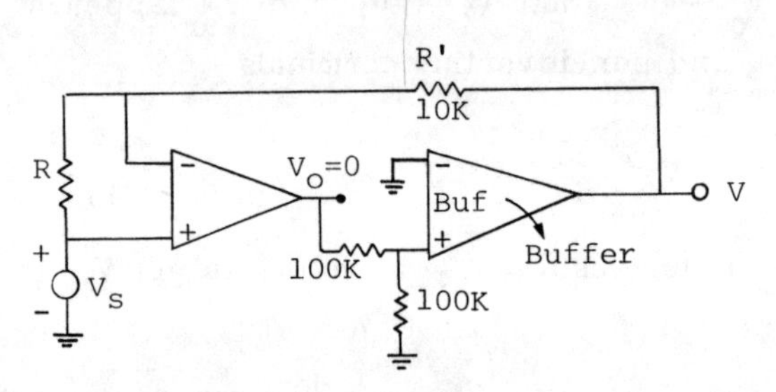

Close S_1 and S_2 ; V' = 0; apply signal v_s.

$$V_o = A_d \cdot V_d + A_c \cdot V_c = 0$$

$$V_1 = V_s \text{ and } V_2 = V_s \cdot \frac{R'}{R+R'} + V \cdot \frac{R}{R+R'} \cong V_s + \frac{V \cdot R}{R'}$$

$$V_d = V_1 - V_2 - V_{io} = \frac{-R}{R'}(V + V_3)$$

$$V_c = V_s + (VR/2R')$$

$$-A_d \frac{R}{R'}(V+V_3) + A_c\left(V_s + \frac{V \cdot R}{2R'}\right) = 0$$

If the measured value of V is V_6, then

$$\rho \cdot \frac{R}{R'}(V_6+V_3) = v_s$$

Slew rate - The slew rate is the maximum rate of change of the output voltage when supplying the rated output.

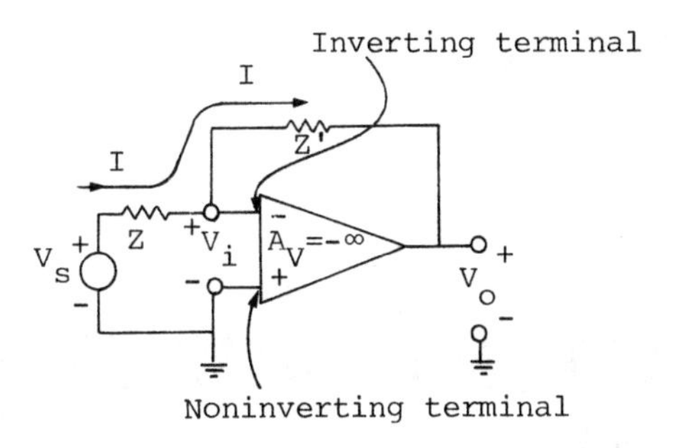

For a single-ended input amplifier, adjust Z = R = 1K and Z' = R' = 10KΩ.

V_s is a high-frequency square-wave. Its slopes are measured with respect to the lines of the leading and trailing edges of the output signal.

The slower of the two is the slew rate.

Frequency response of the op amp:

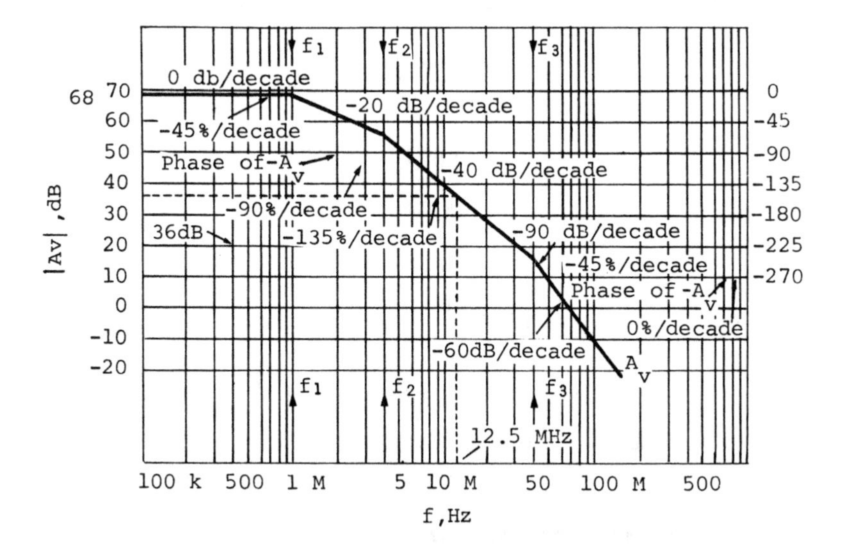

CHAPTER 11

OPERATIONAL AMPLIFIER SYSTEMS

11.1 BASIC OPERATIONAL AMPLIFIER APPLICATIONS

11.1.1 SIGN CHANGER, OR INVERTER

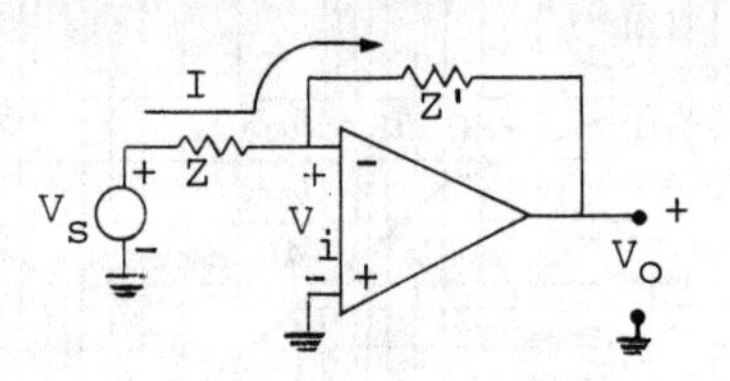

If $Z = Z'$, then $A_{vf} = \frac{V_o}{V_s} = \frac{-Z'}{Z} = -1.$

The sign of the input signal is changed at the output.

The circuit acts as a phase inverter.

11.1.2 SCALE CHANGER

If the ratio $\frac{Z'}{Z} = K$ (a real constant), then $A_{vf} = -K$. The scale has been multiplied by the factor $-K$.

11.1.3 PHASE SHIFTER

If Z and Z' are equal in magnitude and differ in angle, then the op amp shifts the phase of a sinusoidal input voltage.

11.1.4 ADDER

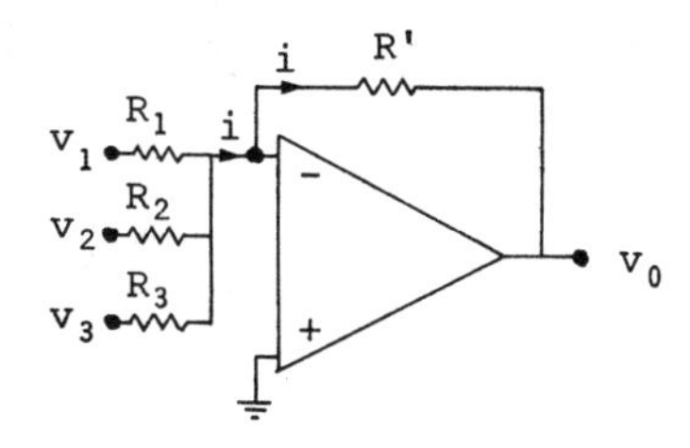

$$v_o = \frac{-R'}{R}(v_1+v_2+v_3), \text{ if } R_1 = R_2 = R_3 = R$$

11.1.5 NON-INVERTING ADDER

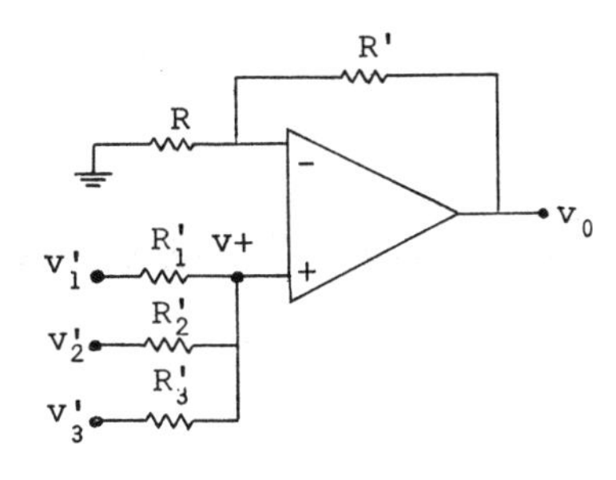

$$v_o = \left(1 + \frac{R'}{R}\right) v_+$$

For n equal to resistors, each of value R'_2

$$\frac{R'_{p_2}}{R'_2 + R'_{p_2}} = \frac{R'_2 \div (n-1)}{R'_2 + [R'_2 \div (n-1)]} = \frac{1}{n}$$

$$v_+ = \frac{1}{n}(v'_1 + v'_2 \ldots).$$

$$R'_{p_2} = R'_1 \,||\, R'_3 \,||\, R'_4 \ldots ||\, R'_n$$

11.1.6 VOLTAGE-TO-CURRENT CONVERTER (TRANSCONDUCTANCE AMPLIFIER)

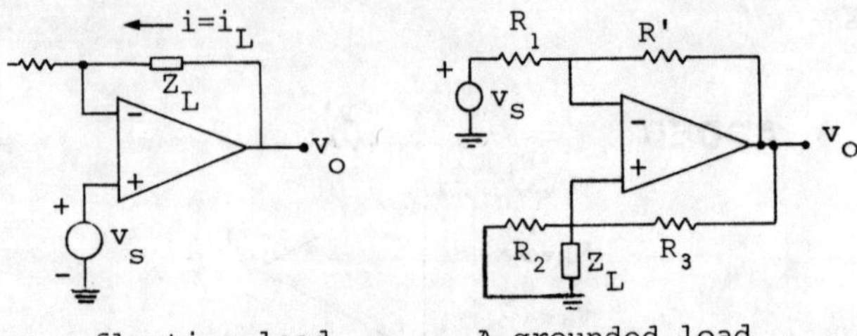

For a floating load A grounded load

Floating load (neither side is grounded):

The current in $Z_L = i_L = \frac{v_{s(t)}}{R_1}$.

Grounded load: $i_L(t) = -\frac{v_{s(t)}}{R_2}$

11.1.7 CURRENT-TO-VOLTAGE CONVERTER

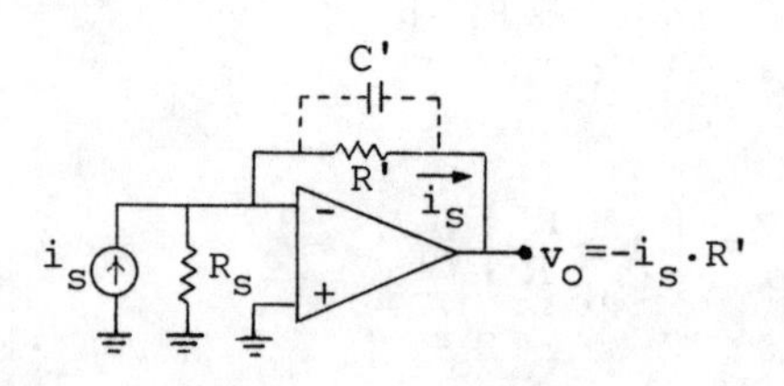

C' is used to reduce the high-frequency noise as well as the possibility of oscillations.

$$v_o = -i_s R'$$

The circuit acts like an ammeter with zero voltage across the meter.

11.1.8 D.C. VOLTAGE FOLLOWER

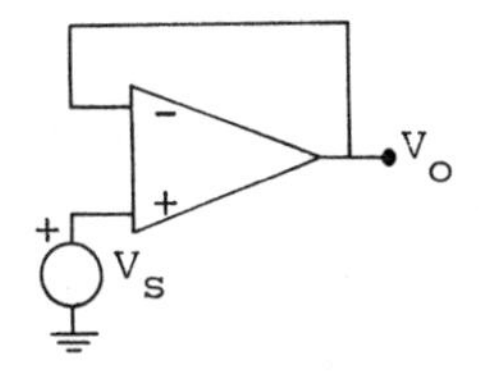

$V_o = V_s$ (because the inputs are tied (virtually) together)

The follower has a high input resistance and a low output resistance.

11.2 AC-COUPLED AMPLIFIER

This is used for amplifying an ac signal, while any dc signal is to be blocked.

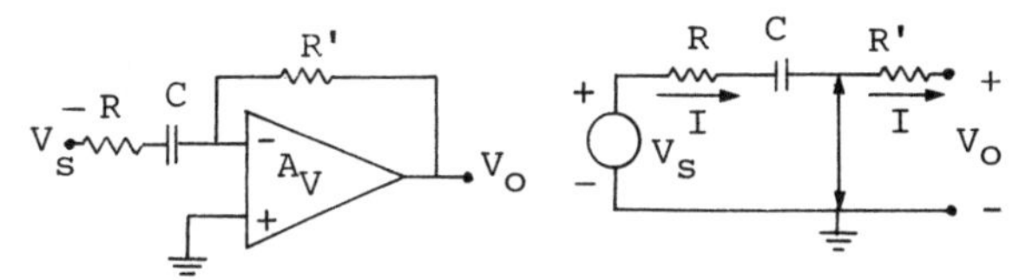

Equivalent circuit

$$V_o = -IR' = \frac{-V_s}{R + \frac{1}{sC}} R'$$

$$A_{vf} = \frac{V_o}{V_s} = \frac{-R'}{R} \quad \frac{s}{s + \frac{1}{RC}}$$

f_L = low 3-dB frequency = $1/2\pi\, RC$

AC voltage follower:

This follower is used to connect a signal source with

high internal source resistance to a load of low impedance, which may be capacitive.

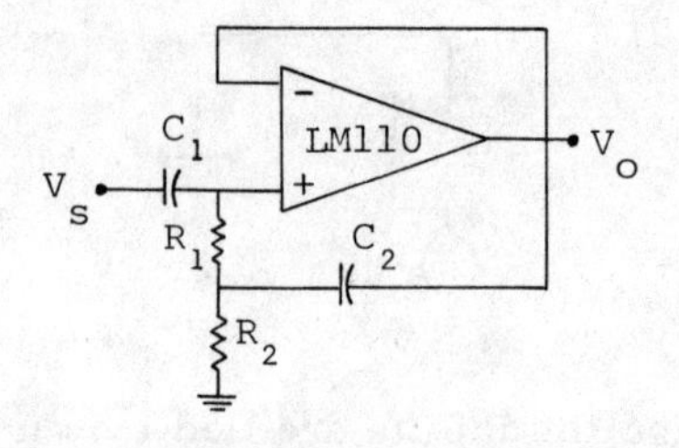

Analog integration and differentiation:

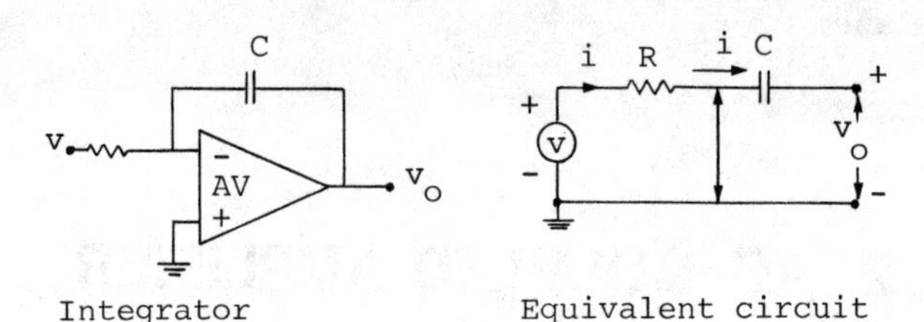

Integrator Equivalent circuit

$$v_o = \frac{-1}{C}\int i dt = \frac{-1}{RC}\int v\,dt$$

If the input voltage is constant, v = V, then the output will be a ramp, $v_o = \frac{-Vt}{RC}$.

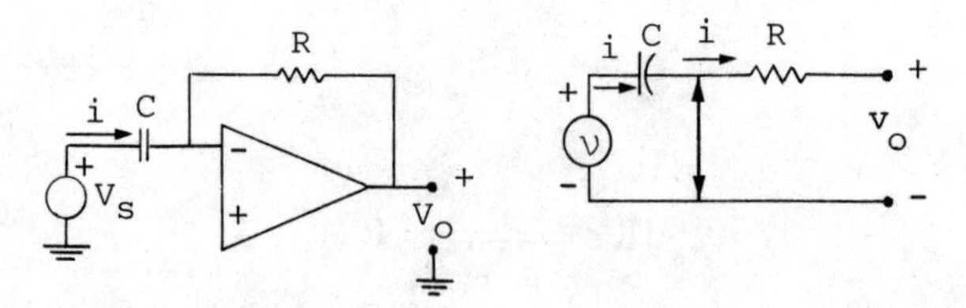

Differentiator Equivalent circuit

$$v_o = -R_i = -RC\,\frac{dv}{dt}$$

If the input signal is $v = \sin \omega t$, then the output will be $v_o = -RC\omega\cos\omega t$. This results in amplification of the high-frequency components of amplifier noise, and the noise output may completely mask the differentiated signal.

11.3 ACTIVE FILTERS

11.3.1 IDEAL FILTERS

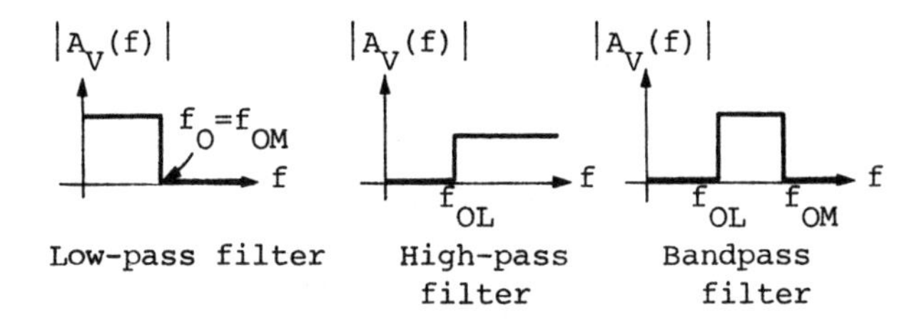

An approximation for an ideal low-pass filter is:

$$\frac{A_V(s)}{A_{VO}} = \frac{1}{P_n(s)}$$

where $P_n(s)$ is a polynomial in the variable s with zeros in the left-hand plane.

11.3.2 BUTTERWORTH FILTER

$P_n(s) = B_n(s)$, known as the "Butterworth polynomial"

$$|B_n(\omega)| = \left[1 + \left(\frac{\omega}{\omega_o}\right)^{2n}\right]^{\frac{1}{2}}$$

Butterworth low-pass filter response:

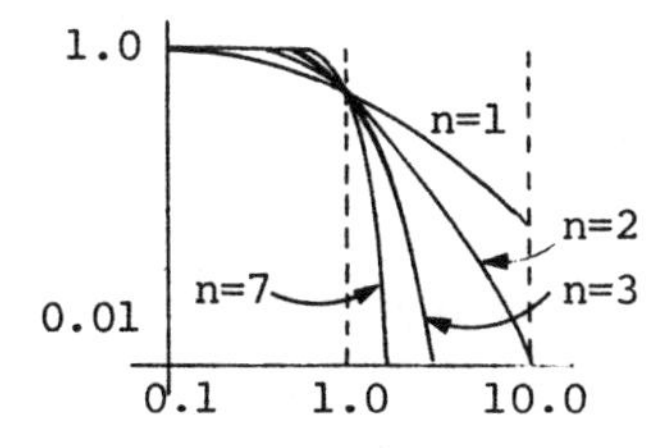

The transfer function is

$$\frac{A_V(s)}{A_{VO}} = \frac{1}{(s/\omega_o)^2+2K(s/\omega_o) + 1}$$

where $\omega_o = 2\pi f_o$ = High-frequency 3-dB point.

This represents a second-order filter.

For the first-order filter, $\frac{A_v(s)}{A_{vo}} = \frac{1}{\frac{s}{\omega_o} + 1}$

Circuit

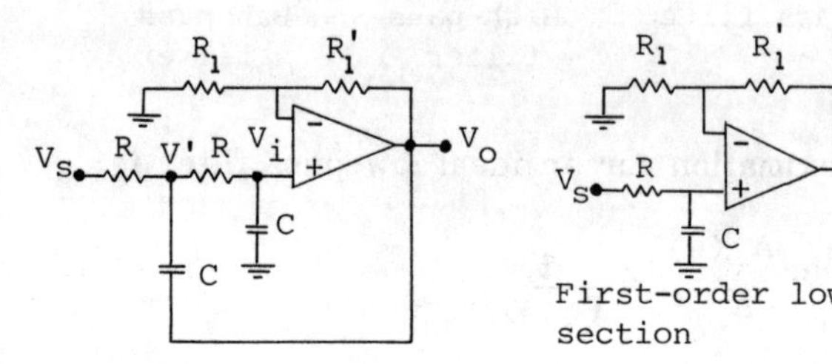

Second-order low-pass section

First-order low-pass section

Mid-band gain of op amp,

$$A_{vo} = \frac{V_o}{V_i} = \frac{R_1 + R_1'}{R_1}$$

$$\frac{A_v(s)}{A_{vo}} = 1/[(RC_s)^2 + (3 - A_{vo})RCs + 1]$$

obtained by applying KCL to node V'.

$$\omega_o = \frac{1}{RC} \quad \text{and} \quad 2K = 3 - A_{vo} \ldots$$

by comparing the coefficients of s^2 in $\frac{A_v(s)}{A_{vo}}$ and typical second-order transfer function.

Even-order Butterworth filters are synthesized by cascading second-orders prototypes such as those shown above and choosing A_{vo} of each op amp such that A_{vo} = 3 - 2K.

Normalized Butterworth polynomials:

n	Factors of $B_n(s)$
1	$(s+1)$
2	$(s^2+1.414\,s+1)$
3	$(s+1)(s^2+s+1)$
4	$(s^2+0.765s+1)(s^2+1.848s+1)$
5	$(s+1)(s^2+0.618s+1)(s^2+1.618s+1)$

Odd-order filters - Cascade the first-order filter with the second order.

CHAPTER 12

FEEDBACK AND FREQUENCY COMPENSATION OF OP AMPS

12.1 BASIC CONCEPTS OF FEEDBACK

Standard inverting configuration:

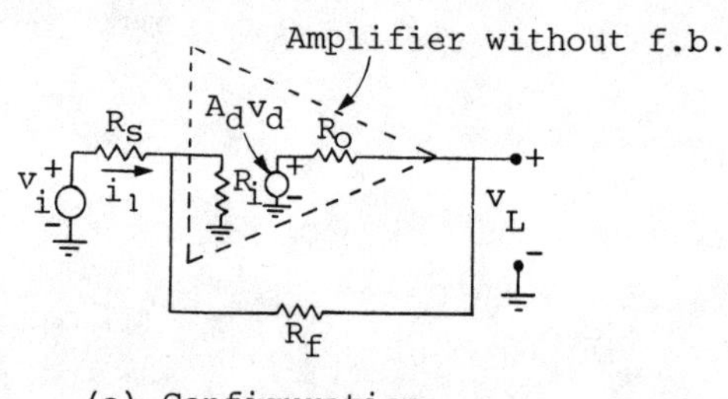

(a) Configuration

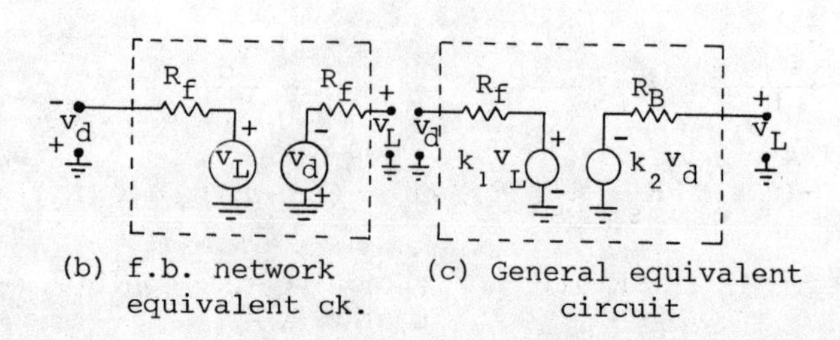

(b) f.b. network equivalent ck.

(c) General equivalent circuit

Current-differencing negative feedback circuit:

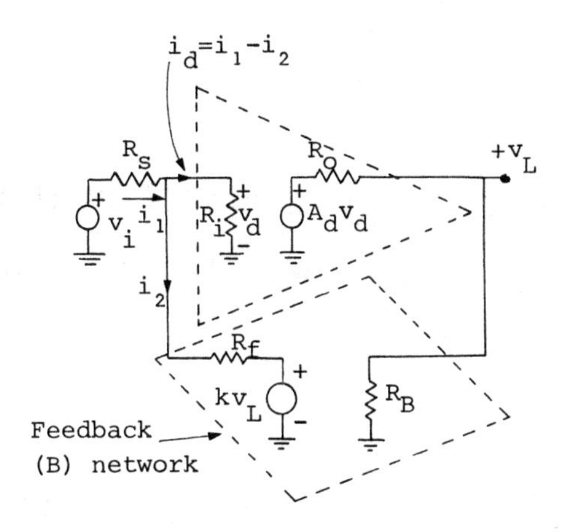

(a) Current-differencing -ve - feedback ckt.

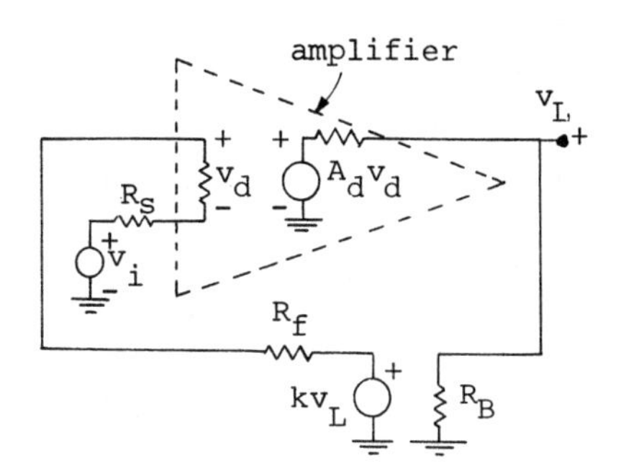

(b) Voltage-differencing -ve f.b. circuit

Gain of a feedback amplifier with current differencing:

$$i_1 = \frac{v_i + v_d}{R_s}, \quad i_2 = \frac{-v_d + K \cdot v_L}{R_f}, \quad i_d = \frac{-v_d}{R_i}$$

$$v_L \cong -(v_i A_d \div R_s)(R_s \parallel R_f) \div [1 + (A_d \cdot K/R_f)(R_s \| R_f)]$$

$$\cong -[R_f \div (K \cdot R_s)]\, v_i \quad \ldots$$ (For inverting configuration $K = 1$)

Loop-gain T of the amplifier:

$$\frac{-A_d K}{R_f}(R_s // R_f)$$

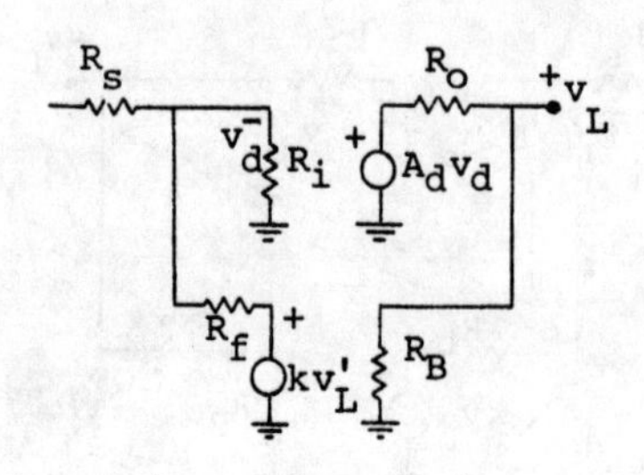

Circuit used to calculate T and A_o:

$$T = \left.\frac{v_L}{v_i'}\right|_{v_i=0} = \frac{v_L}{v_d}\,\frac{v_d}{i_d}\,\frac{i_d}{v_L'}$$

Open loop gain, or gain without feedback A_o:

$$A_o = \left.\frac{v_L}{v_i}\right|_{v_L=0} = \frac{-A_d}{R_s}(R_s \| R_f)$$

Overall gain:

$$A_v = \frac{v_L}{v_i} = \frac{A_o}{1-T}$$

For voltage-differencing circuit:

$$T \cong -A_d \cdot K$$

$$A_o \approx A_d$$

$$v_L \approx v_i A_d/(1+A_d K)$$

The loop gain T:

The loop gain controls the "amount" of feedback present in a circuit.

If $T = 0$ $(K=0)$, there is no feedback.

When $T >> 1$, the gain of an amplifier with feedback approaches $A_v = v_L/v_i = -(R_f/KR_s)$ for the current-differencing configuration, and $A_v = \frac{1}{K}$ for the voltage-differencing configuration. (These amplifiers are shown in the diagram above.) We note that in both cases the gain with feedback is more or less independent of the amplifer gain, A_d. Consequently, the amplifier employing feedback is much more stable against variations in temperature and other parameters as the loop gain increases.

The feedback also has the effect of decreasing the gain from A_o (without feedback) to $\frac{A_o}{(1-T)}$.

Feedback amplifiers and the sensitivity function:

Sensitivity function: $$\int_{A_o}^{A_v} = \frac{dA_v/A_v}{dA_o/A_o}$$

$$\int_{A_o}^{A_v} = 1/(1 - T)$$

12.2 FREQUENCY RESPONSE OF A FEEDBACK AMPLIFIER

Bandwidth and Gain-bandwidth Product:

For current-differencing negative feedback Amplifier:

$$A_v = \frac{A_o}{1 - T}$$

Single-pole amplifier:

$$A_d = \frac{A_{dm}}{1+(s/\omega_1)} \ \ldots \ A_{dm} = \text{gain at low frequency}$$

$$T = \frac{-T_m}{1+\left(\frac{s}{\omega_1}\right)}, \quad T_m = \frac{A_{dm} \cdot K(R_s \| R_f)}{R_f}, \quad A_o = \frac{-A_{om}}{1+\left(\frac{s}{\omega_1}\right)},$$

$$A_{om} = \frac{A_{dm}(R_s \| R_f)}{R_s}$$

$$A_v = \frac{-A_{om}}{1+T_m}\left(\frac{1}{1+\frac{s}{\omega_1}(1+T_m)}\right)$$

A pole is located at $s = -\omega_1(1 + T_m)$

f_n(upper 3-dB frequency) $= f_1(1 + T_m)$

$$G \times (\text{B.W.}) = \left[A_{v(f=0)}\right] \cdot f_n = A_{om} \cdot f_1 \text{(A constant, independent of feedback)}$$

Characteristic:

increasing f.b.
$j\omega$
σ
0
$-\omega_1(1+T_m)$
$-\omega_1$

(a) locus of pole motion

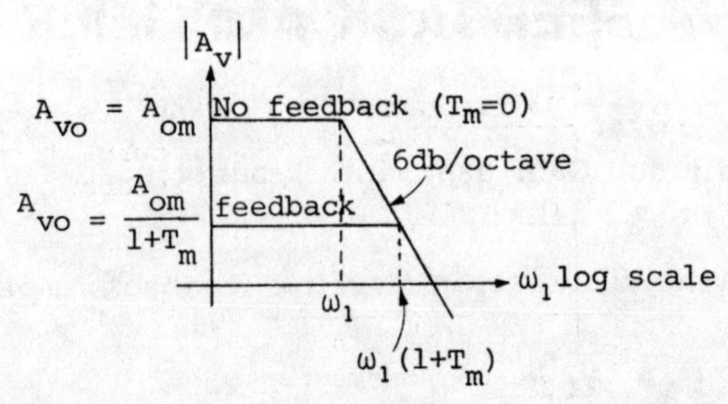

(b) Gain versus frequency.

Double-pole amplifier:

$$A_d(s) = \frac{A_{dm}}{\left(1 + \frac{s}{\omega_1}\right)^2}, \quad T = \frac{-T_m}{\left(1 + \frac{s}{\omega_1}\right)^2}$$

$$A_o = \frac{-A_{om}}{\left(1 + \frac{s}{\omega_1}\right)^2}, \qquad A_v(s) = \frac{-A_{om}}{1 + T_m + (2s/\omega_1) + (s^2/\omega_1^2)}$$

Poles are located at $s = -\omega_n(\xi \pm \sqrt{\xi^2 - 1})$, $\quad \xi = \frac{1}{\sqrt{1 + T_m}}$

Characteristics:

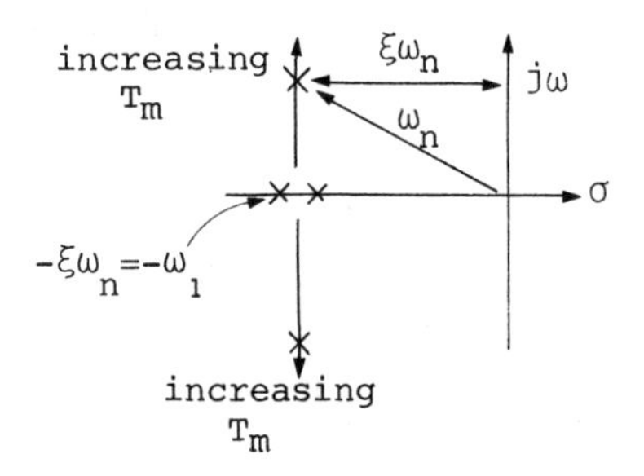

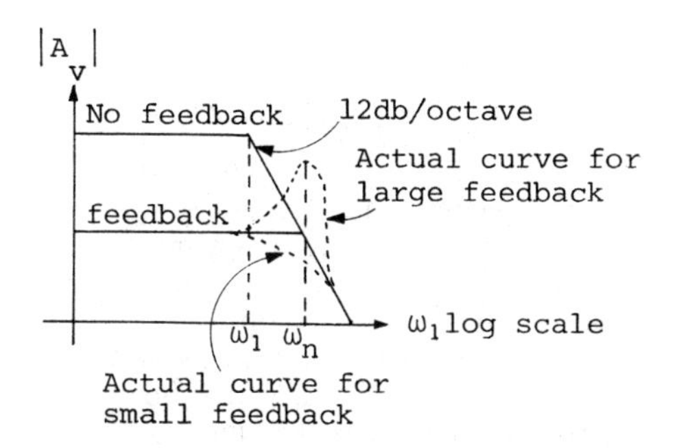

12.3 STABILIZING NETWORKS

No frequency compensation:

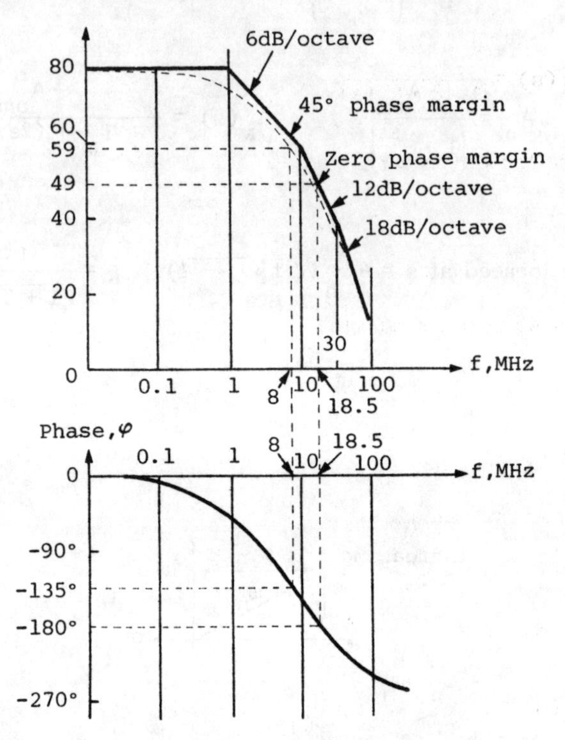

Bode plot for amplifier

The amplifier's open-loop gain is

$$A_o = \frac{-10^4}{\left(1 + \frac{s}{2\pi 10^6}\right)\left(1 + \frac{s}{2\pi \times 10^7}\right)\left(1 + \frac{s}{2\pi \cdot 30 \times 10^6}\right)}$$

If sufficient feedback is applied to make $|T| = 1$ at 18.5 MHZ, the amplifier is said to be marginally stable.

If sufficient feedback is applied to make $|T| = 1$ at a frequency < 18.5MHz, the actual phase at $|T| = 1$ is < 180°. The difference between 180° and the actual phase at $|T| = 1$ is called "Phase-Margin".

The frequency at which $|T| = 1$ is called the gain crossover frequency.

$$T(s) = A_o(s) \cdot K/(R_f/R_s), \quad \beta = \frac{K}{(R_f \div R_s)}$$

$$A_v(s) = \frac{A_o(s)}{1 - \beta \cdot A_o(s)} \approx \frac{-1}{\beta} \text{ (if } |T(s)| >> 1)$$

For a 45° phase margin,

$$|T(8MHz)| = 1 = \beta\,|A_o(8MHz)| \simeq \beta(936), \quad \beta \approx -59dB$$

At low frequencies, $T_m = \beta \cdot A_{om} = 21dB$

For a 45° phase margin:

$$\beta = 1 / |A_o(\omega \text{ corresponding to } 135°)|$$

$$|T_{max}| = \beta|A_{om}| = A_{om} \div |A_o(\omega \text{ corresponding to } 135°)|$$

Simple lag compensation:

Simple lag compensation is designed to introduce an additional negative real pole in the transfer function of the open-loop amplifier gain, A_o.

$$A_{o1}(s) = A_o(s) \div \left(1 + \frac{s}{\alpha}\right) \ldots.$$

The pole α is adjusted so that $|T|$ drops to OdB at a frequency where the poles of A_o contribute negligible phase shift.

This pole increases T but decreases the cross-over frequency.

Effect of lag compensation:

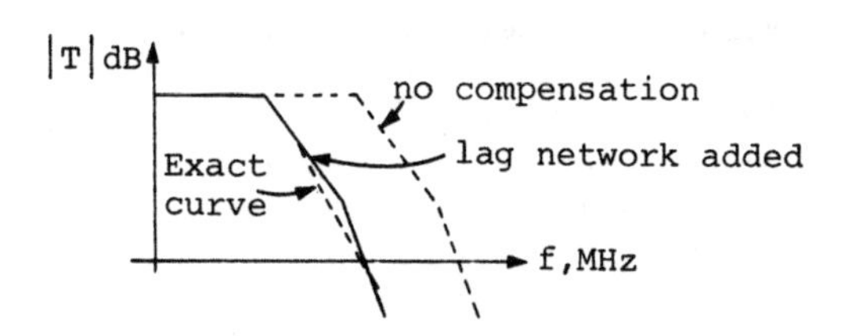

Lag compensating network:

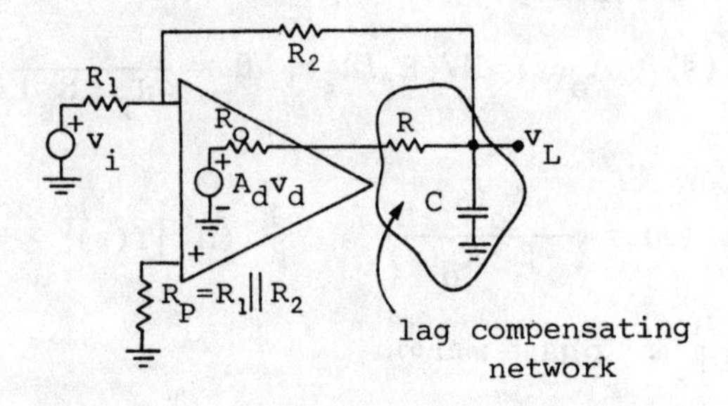

The break frequency of this lag filter:

$$\frac{\alpha}{2\pi} \approx \frac{1}{2\pi(R+R_o)C} \quad \ldots \text{ (assuming } R+R_o << R_2\text{)}.$$

Lead compensation:

The transfer function $H(s) = \dfrac{s+\delta_1}{s+\delta_2}$

($\delta_2 > \delta_1$, i.e., the pole is located at a higher frequency than the zero.)

Bode diagram for H(s) and the network:

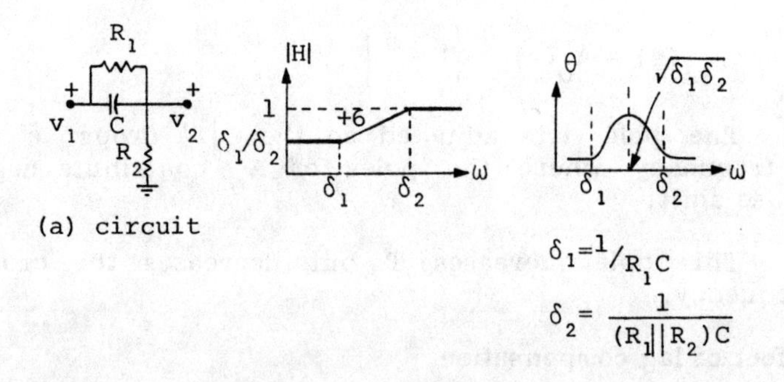

This network introduces a low-frequency attenuation:

$$H(0) = R_2 \div (R_1+R_2) = \delta_1/\delta_2$$

CHAPTER 13

MULTIVIBRATORS

13.1 COLLECTOR-COUPLED MONOSTABLE MULTIVIBRATORS

Stable-state:

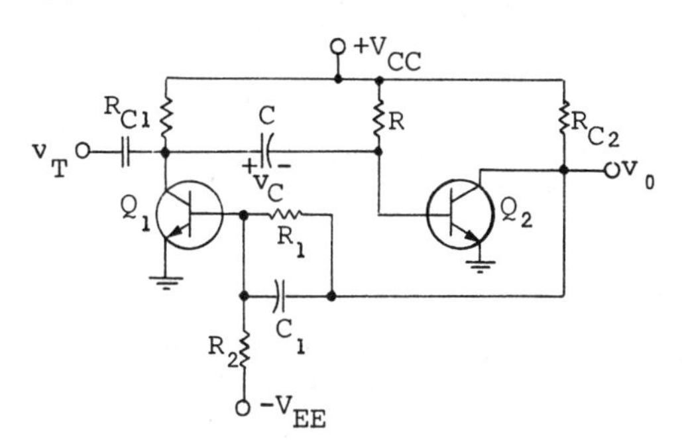

Collector-coupled monostable

In this state Q_1 is off, Q_2 is saturated and v_o is low.

$v_{c_1} \approx V_{cc}$ and $v_{c_2} \approx 0$, the current in R must be sufficient to saturate Q_2, i.e.,

$$I_{B_2} = I_R = \frac{(V_{cc} - V_{BES_2})}{R} \gg I_{BS_2} = \frac{I_{cs_2}}{\beta}$$

or

$$\frac{V_{cc} - V_{BES_2}}{R} \geq \frac{V_{cc} - V_{CES_2}}{\beta_2 \cdot R_{c_2}}$$

C_1 is a speed-up capacitor to couple change in v_o to v_{B_1}, turning Q_1 on rapidly.

Circuit operation:

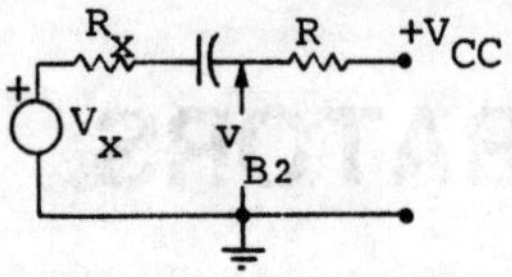

Voltages for the monostable shown in the figure in steady-state and just after triggering into the Quasi-Stable by a trigger pulse at t = 0.

	v_{C_1}	v_{B_2}	v_o	v_{B_1}	Q_1	Q_2
Stable state (t=0)	$\simeq V_{cc}$	$\simeq 0V$	$\simeq 0V$	$< 0V$	off	on
Quasi-stable state ($t=0^+$)	$\simeq 0V$	$\simeq -V_{cc}$	$\simeq V_{cc}$	$\simeq 0V$	on	off

The initial trigger pulse need not be of amplitude V_{cc}. It only starts Q_2 off; the circuit's regenerative action itself drives v_{c_1} to $\approx$ 0v and v_{B_2} to $\approx -V_{cc}$.

Output pulse width:

The equivalent-circuit at the collector of Q_1 at $t = 0^+$ when Q_1 has just turned on and Q_2 has just turned off.

τ (Time constant with which v_{B_2} moves toward V_{cc})

$$= (R+R_x) \cdot C \approx R \cdot C, \ (R >> r_{sat})$$

$$v_{B_2}(t) = V_F - (V_F - V_I)e^{-t/\tau}$$

(where $V_1 = V_{BE_2} - V_{cc} + V_{CES_1}$

= initial value of v_{B_2}, and

$V_F = V_{cc}$ = the final value of V_{B_2})

T = The duration of the output pulse

$$= \tau \cdot \ln \frac{2V_{cc} - V_{BES_2} - V_{CES_1}}{V_{cc} - V_{BET_2}}$$

For a typical transistor, $V_{BES} + V_{CES} \approx 2V_{BET}$

$$T \approx \tau \ln 2 \approx 0.69\tau \approx 0.69R_c$$

Recovery time:

This is the time one must wait before the monostable multivibrator should be triggered.

When the quasi-stable state ends at t = T, the output returns low although the circuit has not returned to its stable state.

At $t = T^-$, $v_{c_1} = V_{CES_1}$, $v_{B_2} = V_{BET_2}$ and the voltage across C is

$$v_c = V_{CES_1} - V_{BET_2} \approx 0.$$

The v_{c_1} increases exponentially from $v_I = V_{CES}$ towards its final value, $V_f = V_{cc}$, with

$$\tau_R \text{ (Recovery time constant)} = (R_{c_1} + (R \| R_{i_2}) \cdot C$$

$$\approx (R_{c_1} + R_{i_2}) \cdot C$$

$$V_{cc}' \text{ (the output-volt with } v_{EE} = 0) = \frac{V_{cc} \cdot R_1 + V_{BES_1} \cdot R_{c_2}}{R_1 + R_{c_2}}$$

Waveforms:

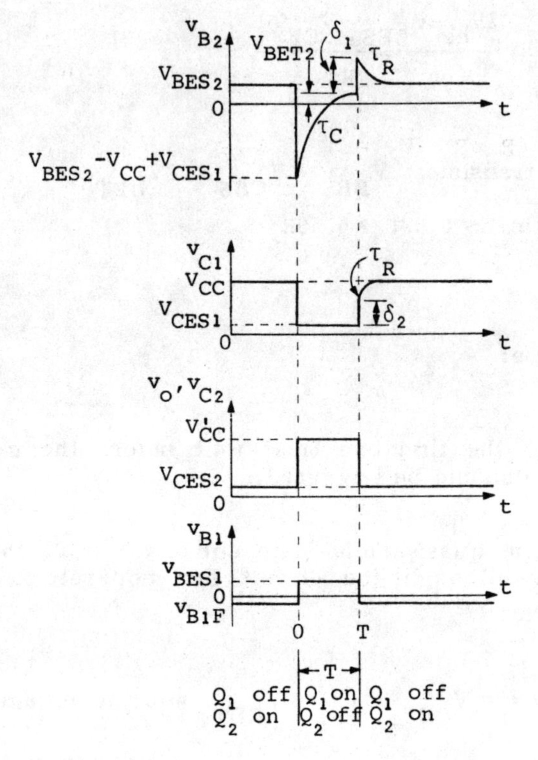

Waveforms for the collector-coupled monostable

13.2 EMITTER-COUPLED MONOSTABLE MULTIVIBRATORS

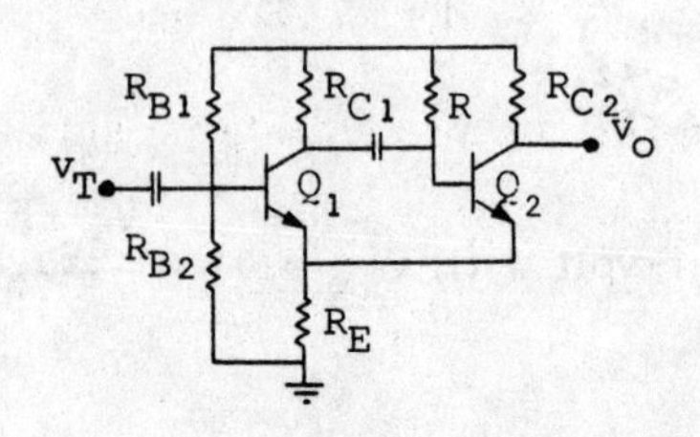

Circuit operation:

$$(R+R_E)I_{B_2} + R_E \cdot I_{c_2} = V_{cc} - V_{BES_2}$$ (KVL equation for Q_2 with Q_1 off and Q_2 on)

$$R_E \cdot I_{B_2} + (R_E + R_{c_2})I_{c_2} = V_{cc} - V_{CE_2}$$

$$v(0) = V_{cc} - R_{c_2}(V_{cc} - V_{CES_2})/(R_{c_2} + R_E)$$

$$= R_E(V_{cc} - V_{CES_2})/(R_{c_2} + R_E) + V_{CES_2}$$

$$v_E(0^-) = v_o(0^-) - V_{CES_2}$$

$$v_{B_2}(\bar{0}) = v_E(0) + V_{BES_2}$$

$$v_{B_1}(\bar{0}) < v_E(\bar{0}) + V_{BET_1}$$ (To maintain Q_1 off)

$$v_{c_1}(\bar{0}) = v_{cc}$$ (with Q_1 off)

$$v_{B_1}(0^+) = V_{B_1} = \frac{V_{cc} \cdot R_{B_2}}{R_{B_1} + R_{B_2}} = v_{B_1}(\bar{0})$$

$$v_E(0^+) = V_{B_1} - V_{BES_1}$$

$$i_{c_1}(0^+) \approx i_{E_1}(0^+) = v_E(\bar{0})/R_E, \quad v_{c_1}(0^+) = V_{cc} - i_{c_1}(0^+)R_{c_1}$$

The condition for Q_1 in a quasi-stable state:

$$\frac{R_E + R_{c_1}}{R_E} > \frac{R_{B_1} + R_{B_2}}{R_{B_2}}$$

Voltages with a Narrow Trigger Pulse Applied at t = 0

Parameter	Stable state $t = 0^-$ (Q_1 off, Q_2 saturated)	Quasi-stable state $t = 0^+$ (Q_1 saturated, Q_2 off)
v_0	$V_{CC} - (V_{CC} - V_{CES_2}) \frac{R_{C_2}}{R_E + R_{C_2}}$ $\cong V_{CC}\left(1 - \frac{R_{C_2}}{R_E + R_{C_2}}\right)$	V_{CC}
v_E	$v_0(0^-) - V_{CES_2}$ $= (V_{CC} - V_{CES_2})\left(1 - \frac{R_{C_2}}{R_E + R_{C_2}}\right)$ $\cong V_{CC}\left(1 - \frac{R_{C_2}}{R_E + R_{C_2}}\right)$	$V_{B_1} - V_{BES_1}$ $\cong V_{CC} \frac{R_{B_2}}{R_{B_1} + R_{B_2}}$
v_{B_2}	$v_E(0^-) + V_{BES_2}$ $= (V_{CC} - V_{CES_2})\left(1 - \frac{R_{C_2}}{R_E + R_{C_2}}\right) + V_{BES_2}$ $\cong V_{CC}\left(1 - \frac{R_{C_2}}{R_E + R_{C_2}}\right)$	$v_{B_2}(0^-) + \left[v_{C_1}(0^+) - v_{C_1}(0^-)\right]$ $\cong V_{CC}\left(\frac{R_{B_2}}{R_{B_1} + R_{B_2}} - \frac{R_{C_2}}{R_E + R_{C_2}}\right)$
v_{B_1}	$\cong V_{CC} \frac{R_{B_2}}{R_{B_1} + R_{B_2}}$	$\cong V_{CC} \frac{R_{B_2}}{R_{B_1} + R_{B_2}}$
v_{C_1}	V_{CC}	$v_E(0^+) + V_{CES_1}$ $\cong \frac{V_{CC} R_{B_2}}{R_{B_1} + R_{B_2}}$

Condition for operation:

Conditions for Operation of the Monostable

1) $v_{B_1}(0^-) < v_E(0^-) + V_{BET_1}$

2) $i_{B_2}(0^-) > i_{C_2}(0^-)/\beta$

3) $v_{C_1}(0^+) < v_{B_1}(0^+)$

4) $v_{B_2}(0^+) < v_E(0^+) + V_{BET_2}$

Waveforms - With Q_1 saturated and Q_2 off in the quasi-stable state and Q_1 off and Q_2 saturated in the stable state.

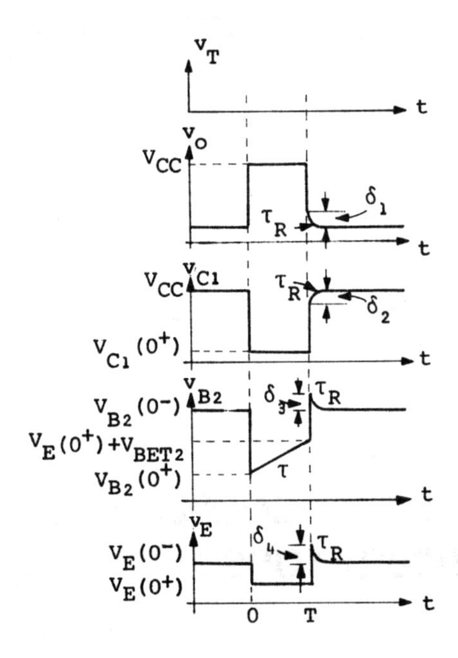

Pulse width and recovery:

$$T = \tau \cdot \ln\left[\frac{V_F - V_I}{V_F - (V_E(0+) + V_{BET_2})}\right]$$

$$\approx R \cdot C \ln \left[\frac{1 + \frac{R_{C_2}}{R_E + R_{C_2}} - \frac{R_{B_2}}{R_{B_1} + R_{B_2}}}{1 - \frac{R_{B_2}}{R_{B_1} + R_{B_2}}} \right]$$

$$\tau_R = \text{(Recovery time constant)}$$

$$= (R_{C_1} + \lambda_{\pi_2} + R_E \parallel R_{C_2})C$$

τ_R determines the circuits retrigger rate and the rate of decay of the overshoot.

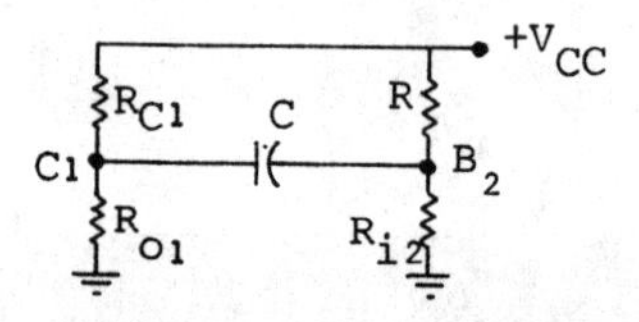

Model of an emitter-coupled monostable multivibrator during the recovery time t > T.

The recovery time for the emitter-coupled monostable multivibrator is greater than in the collector-coupled monostable multivibrator.

13.3 COLLECTOR-COUPLED ASTABLE MULTIVIBRATORS

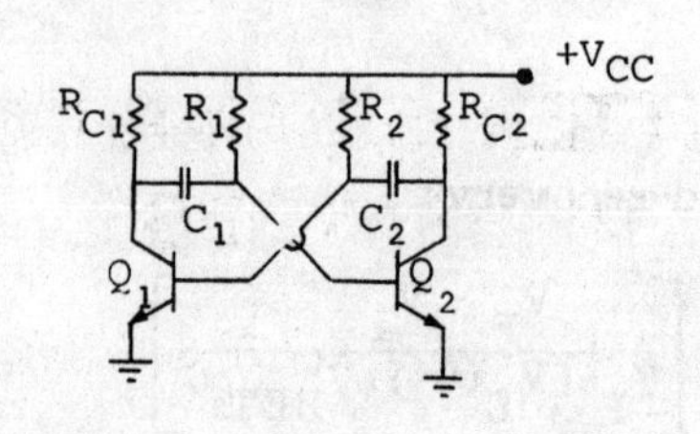

Typical waveforms at each collector and base:

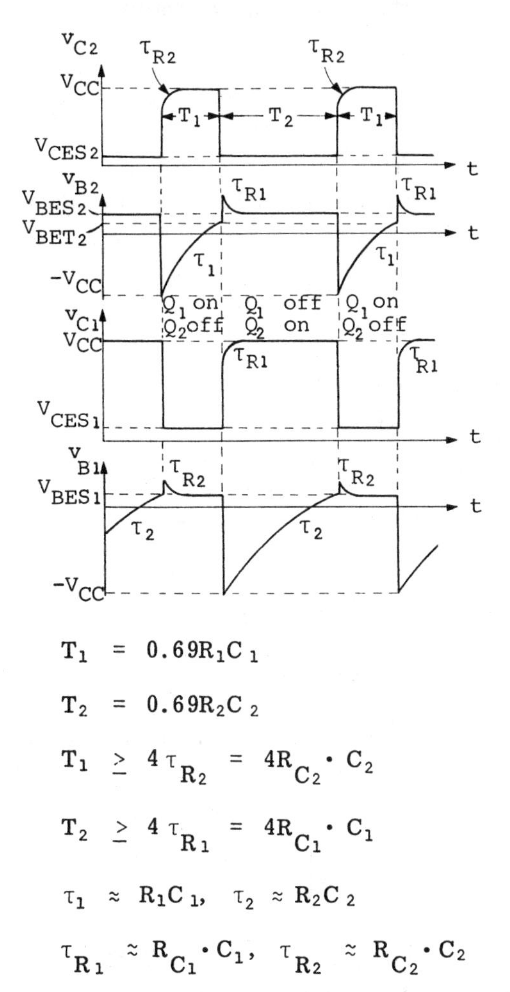

$$T_1 = 0.69R_1C_1$$

$$T_2 = 0.69R_2C_2$$

$$T_1 \geq 4\tau_{R_2} = 4R_{C_2} \cdot C_2$$

$$T_2 \geq 4\tau_{R_1} = 4R_{C_1} \cdot C_1$$

$$\tau_1 \approx R_1C_1, \quad \tau_2 \approx R_2C_2$$

$$\tau_{R_1} \approx R_{C_1} \cdot C_1, \quad \tau_{R_2} \approx R_{C_2} \cdot C_2$$

13.4 EMITTER-COUPLED ASTABLE MULTIVIBRATORS

When the bias levels in the emitter-coupled monostable multivibrator are properly adjusted, the circuit becomes astable. Only one capacitor controls the timing of Q_1 and Q_2.

To keep Q_2 active with c open, require

$$\beta \cdot i_{B_2} < i_{CS_2} \text{ or } \beta \cdot \frac{(V_{CC} - V_{B_2})}{R} < \frac{V_{CC} - v_{C_2}}{R_{C_2}},$$

which, with $v_{B_2} = v_{C_2}$, reduces to $\beta \cdot R_{C_2} < R$.

To keep Q_1 active with c open, V_{B_1} should fall inside the following interval:

$$V_{BE} + \frac{R_E(1+\beta)(V_{CC} - V_{BE})}{R+(1+\beta)R_E} < V_{B_1}$$

$$< \frac{V_{CC} \cdot R_E[R+(1+\beta)R_{C_1}] + V_{BE} \cdot R \cdot R_{C_1}}{R(R_{C_1} + R_E) + (1+\beta)R_E \cdot R_{C_1}}$$

$$T_1 = RC \ln\left[\frac{V_{CC} - v_{B_2}(0^+)}{V_{CC} - V_{B_1}}\right]$$

$$\tau_{R_1} = (R_{C_1} + R \| R_{C_2} \| R_E)C \approx (R_{C_1} + R_{C_2} \| R_E)C$$

$$\tau_{R_2} = [(1+\beta)R_E \| R]C$$

Waveforms:

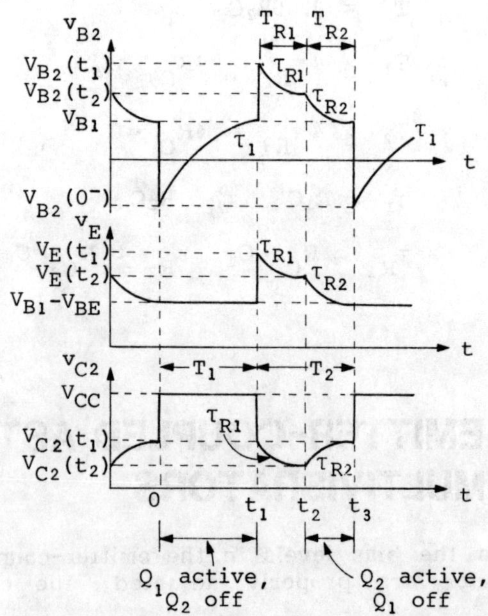

Approximate Voltages

	$t = 0^-$ Q_1 off Q_2 active	$t = 0^+$ Q_1 active Q_2 off
v_{C_1}	V_{CC}	$V_{CC} - v_E(0^+)R_{C_1}/R_E$
v_{B_2}	V_{B_1}	$V_{B_1} - v_E(0^+)R_{C_1}/R_E$
v_E	$V_{B_1} - V_{BE_2}$	$V_{B_1} - V_{BE_1}$

CHAPTER 14

LOGIC GATES AND FAMILIES

14.1 LOGIC LEVEL CONCEPTS

"1" and "0" limits must be separated by some voltage range to ensure that the state of a line can be determined even in the presence of noise, etc.

Voltages	G_e V	S_i V
V_{BET}	0.1	0.5
V_{BES}	0.3	0.7
V_{CES}	1.0	0.1
V_{BE} (active)	0.2	0.6

V_{BET}: The v_{BE} threshold voltage below which little conduction occurs.

V_{BES} and V_{CES}: The v_{BE} and v_{CE} values in saturation

A high or "1" level corresponding to a voltage above some minimum upper level and a low or "0" to a voltage below some maximum lower level.

Graphical interpretation of V_{BET}, V_{BES} and V_{CES}:

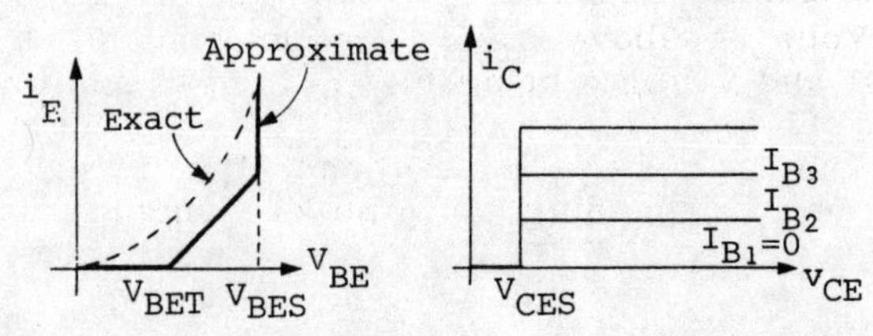

Demonstration of the logic level concept:

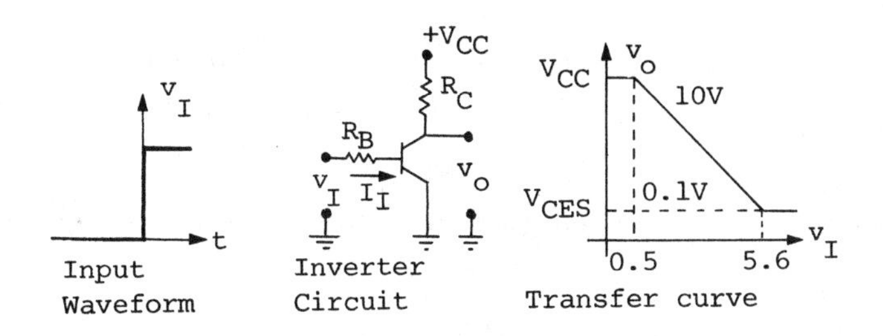

$I_{CS} = (V_{CC} - V_{CES}) \div R_C$ ....In saturation, with no load

$I_{BS} = I_{CS}/\beta$ ....Corresponding base current of the edge of saturation.

$V_I = [R_B(V_{CC} - V_{CES}) \div (\beta RC)] + V_{BES}$ ....Voltage to saturate a transistor

The circuit will saturate for any

$$V_I \geq [R_B(V_{CC} - V_{CES}) \div (\beta R_C)] + V_{BES} = 5.65$$

...[Assuming V_{CC} = +10V, R_C = 1K, R_B = 10K]

With v_I satisfying the above equation, $V_o = V_{CES} = 0.1V$.

Inverter logic levels for Fig. 2.2:

v_I, Volts	v_o, Volts
$\geq$ 5.65	0.1
$\leq$ 0.5	10

V	Level
5.65 to 10V	"1"
-2 to 0.5V	"0"

Voltages above 5.65V are designated as "high" or "1" levels; and voltages below 0.5V, as "low" or "0" levels.

The upper "1" limit and the lower "0" limit are arbitrary, depending on power supply and breakdown voltage.

Logic level notation:

Parameter	Definition
V_{OL}	Low or "0" level output voltage
V_{OH}	High or "1" level output voltage
V_{IL}	Low-level input voltage
V_{IH}	High-level input voltage
$\overline{V}_{IL}$	Maximum V_{IL} level
$\overline{V}_{IH}$	Minimum V_{IH} level
I_{IH}, I_{IL}	Input current for $V_I = V_{OH}$ and $V_I = V_{OL}$, respectively.
I_{OH}, I_{OL}	Maximum available output load current for $V_O = V_{OH}$ and $V_O = V_{OL}$, respectively.

14.2 BASIC PASSIVE LOGIC

14.2.1 RESISTOR LOGIC (RL)

AND function can be implemented with passive components only, such as resistors.

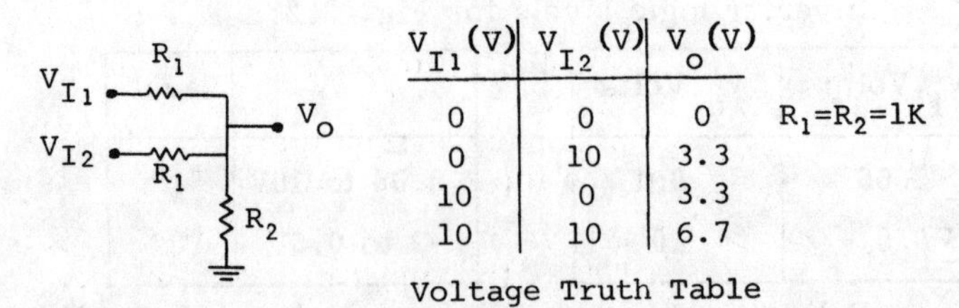

V_{I1} (V)	V_{I2} (V)	V_o (V)
0	0	0
0	10	3.3
10	0	3.3
10	10	6.7

Voltage Truth Table

Fan-out (FO):

The number of equivalent gate inputs that can be driven from the output of a similar gate is the fan-out.

Noise Margin (NM):

This is a measure of the amount of noise that can be tolerated on signal lines before the voltage levels on these lines cease to look like "1s" or "0s".

Graphical Interpretation of NM and "1" and "0":

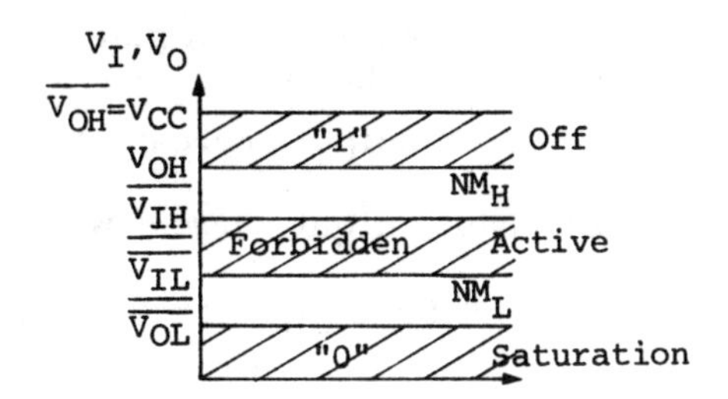

14.2.2 DIODE LOGIC CIRCUITS

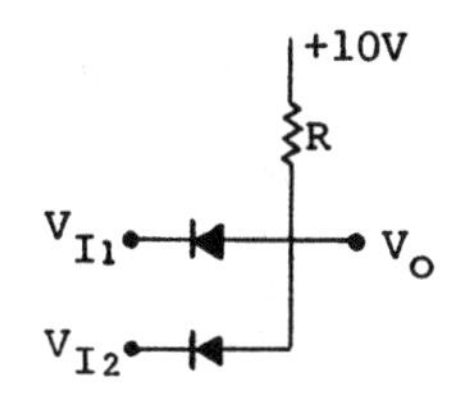

Basic Circuit

V_{I1} V	V_{I2} V	V_O V
0	0	0.7
0	5	0.7
5	0	0.7
5	5	5.7

14.2.3 LOGIC CIRCUIT CURRENT, VOLTAGE, AND PARAMETER DEFINITIONS AND NOTATION

Parameter	Definition and test condition
$\underline{V_{IH}}$	Minimum input voltage that will look like a "1" at the input of a gate in worst case.
$\overline{V_{IL}}$	Maximum input voltage that will still look like a "0" at the input of a gate in worst case.
$\underline{V_{OH}}$	Minimum output voltage which, when applied to $FO_H = N_H$ other gate inputs, will look like a "1" at the input of each driven gate with a NM_H.

$\overline{V_{OL}}$ Maximum output voltage which, when applied to $FO_L = N_L$ other gate inputs, will look like a "0" at the input of each driven gate with a NM_L.

I_{IH} Current required at a gate input with $\underline{V_{OH}}$ present at the input.

I_{IL} Current required at a gate input with $\overline{V_{OL}}$ present at the input.

I_{OH} Available output current when $V_o = \underline{V_{OH}}$.

I_{OL} Available output current when $V_o = \overline{V_{OL}}$.

$FO_H = |I_{OH}/I_{IH}|$ Fan-out for a high-level output.

$FO_L = |I_{OL}/I_{IL}|$ Fan-out for a low-level output.

FO The smaller of FO_H and FO_L.

$NM_H = \underline{V_{OH}} - \underline{V_{IH}}$ Noise margin for a high-level output.

$NM_L = \overline{V_{IL}} - \overline{V_{OL}}$ Noise margin for a low-level output.

NM The smaller of NM_H and NM_L.

14.3 BASIC ACTIVE LOGIC

14.3.1 RESISTOR-TRANSISTOR LOGIC (RTL)

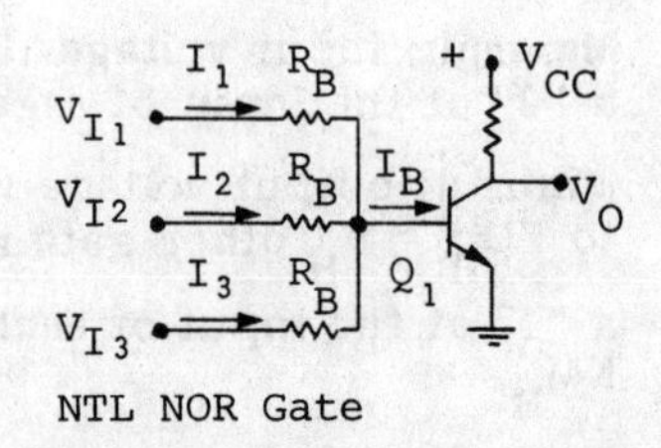

NTL NOR Gate

RTL parameter expressions:

1) $\underline{V_{IH}} = \frac{R_B}{\beta}\left(\frac{V_{CC}-V_{CES}}{R_C} + FO_L|I_{IL}|\right) + 3V_{BES} - 2\overline{V_{OL}}$

2) $\underline{V_{OH}} = NM_H + \underline{V_{IH}}$

3) $\underline{V_{OH}} = (V_{CC}R_B + FO_H V_{BES} R_C)/(R_B + FO_H R_C)$

4) $|I_{OH}| = (V_{CC} - \underline{V_{OH}})/R_C$

5) $I_{IH} = (\underline{V_{OH}} - V_{BES})/R_B$

6) $FO_H = |I_{OH}/I_{IH}|$

7) $\overline{V_{IL}} = V_{BET}$

8) $\overline{V_{OL}} = \overline{V_{IL}} - NM_L = V_{CES}$

9) $I_{IL} = (\overline{V_{OL}} - V_{BES})/R_B$

10) $I_{OL} = \beta[\underline{V_{IH}} - V_{BES} - 2(V_{BES} - \overline{V_{OL}})]/R_B - (V_{CC}-V_{CES})/R_C$

11) $FO_L = |I_{OL}/I_{IL}|$

14.3.2 DIODE-TRANSISTOR LOGIC (DTL)

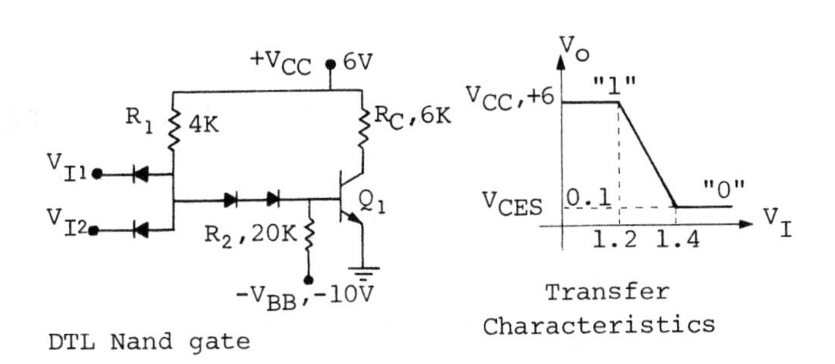

DTL Nand gate

Transfer Characteristics

DTL NAND gate characteristics:

Table 1	Table 2
$\underline{V_{OH}} = 6V$	$FO_L = FO = 9$
$\overline{V_{OL}} = .412V$	$NM_L = NM = .788V$
$\underline{V_{IH}} = 1.4V$	$I_{IL} = -0.83mA$
$\overline{V_{IL}} = 1.2V$	$I_{OL} = 12.2mA$

DTL NAND gate apramaters:

$$\underline{V_{IH}} = V_{DX} + V_{DY} + V_{BES} - V_{D_1}, \quad \overline{V_{IL}} = V_{DX} + V_{DY} + V_{BET} - V_{D_1}$$

$$\left|I_{IL}\right| = (V_{CC}-V_D-V_{CES})/R_1 - (V_{CES}+V_D-V_{DX}-V_{DY}+V_{BB})/R_2$$

$$I_{OL} = \frac{\beta(V_{CC}-2.1)}{R_1} - \frac{\beta(V_{BES}+V_{BB})}{R_2} - \frac{V_{CC}-V_{CES}}{R_C}$$

$$V_{OL} = V_{CES} + \left[\frac{V_{CC}-V_{OL}}{R_C} + (FO_L)\left|I_{IL}\right|\right] r_{sat},$$

$$\overline{V_{OL}} = \overline{V_{IL}} - NM_L$$

14.4 ADVANCED ACTIVE LOGIC GATES

14.4.1 TRANSISTOR-TRANSISTOR LOGIC (TTL)

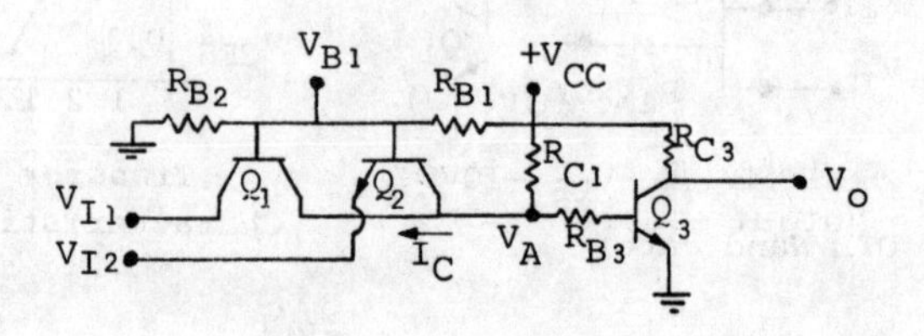

V_{CC}, R_{c_1} and R_{B_3} are chosen so that with Q_1 and Q_2 off, the current from V_{CC} through R_{C_1} is sufficient to saturate Q_3:

$$\frac{V_{CC} - V_{BES_3}}{R_{C_1} + R_{B_3}} = I_{B_3} \geq I_{BS_3}$$

$$= \frac{V_{CC} - V_{CES}}{R_{C_3}} + FO_L I_{IL}$$

Q_1 and Q_2 are off when both inputs V_{I_1}, and V_{I_2}, are low. Furthermore, Q_3 will be turned of if Q_1 or Q_2 , or both are turned on when V_{I_1} or V_{I_2}, or both, are high. Hence, this circuit acts as a NAND gate.

Totem-pole output with phase-splitter driver:

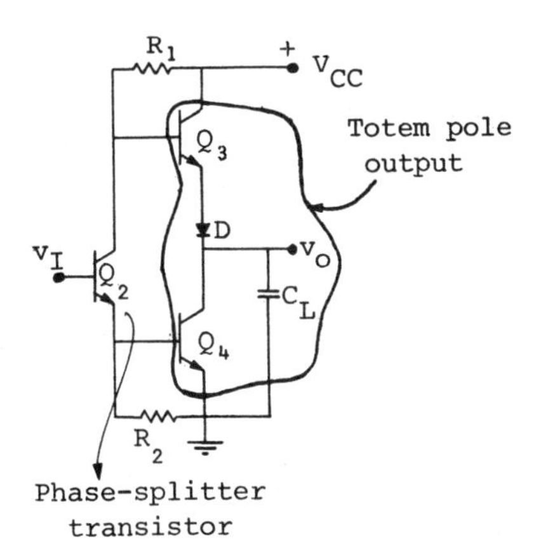

The circuit's response to a positive pulse at the base of Q_2: With v_I low, Q_2 is off, V_{C_2} is high and Q_3 is on, while Q_4 is off and v_o is high.

When v_I goes positive, Q_2 saturates and V_{C_2} droops, turning Q_3 off. V_{E_2} then rises and Q_4 saturates. $(\beta+1)I_{B_4}$ output current quickly discharges C_L and v_o rapidly falls to V_{CES}. When v_I drops from a high level to

OV, Q_2 turns off. V_{E_2} falls to OV, turning Q_4 off; and V_{C_2} rises, turning on Q_3. Q_3 operates as an EF and quickly charges C_L with the large available current $(\beta+1)I_{B_3}$ and v_o rapidly rises.

The totem-pole output thus utilizes EF Q_3 to rapidly raise v_o and the CE transistor Q_4 to rapidly discharge v_o. The phase splitter Q_2 provides the proper phase drive for the totem-pole output transistors. Other topologies, such as DTL, can also use this output connection.

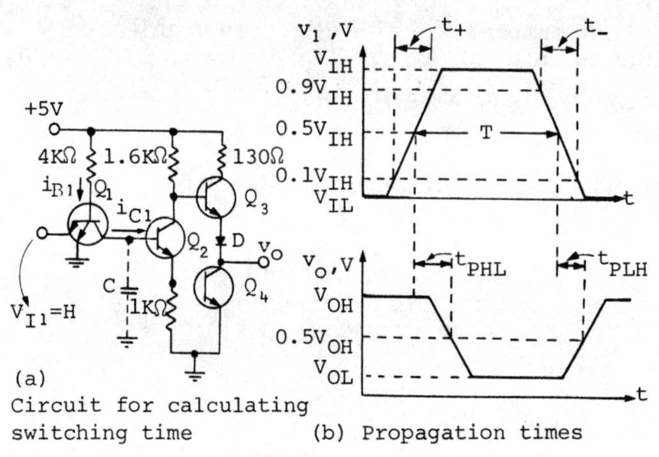

(a) Circuit for calculating switching time

(b) Propagation times

SWITCHING AND PROPAGATION TIMES

Characteristics:

Guaranteed Input and Output Voltage Levels for TTL

Parameter	Value, V	Interpretation for a TTL NAND gate
$\overline{V_{IL}}$	0.8	An input voltage $\le$ 0.8V is guaranteed to turn on Q_1(E-B junction forward biased).
$\underline{V_{IH}}$	2.0	An input voltage $\ge$ 2.0V is guaranteed to turn off Q_1(E-B junction reverse biased).
$\overline{V_{OL}}$	0.4	With $V_I \le \overline{V_{IL}}$, the output is guaranteed to be $\le$ 0.4V under full fan-out.
$\underline{V_{OH}}$	2.4	With $V_I \ge \underline{V_{IH}}$, the output is guaranteed to be $\ge$ 2.4V under full fan-out.

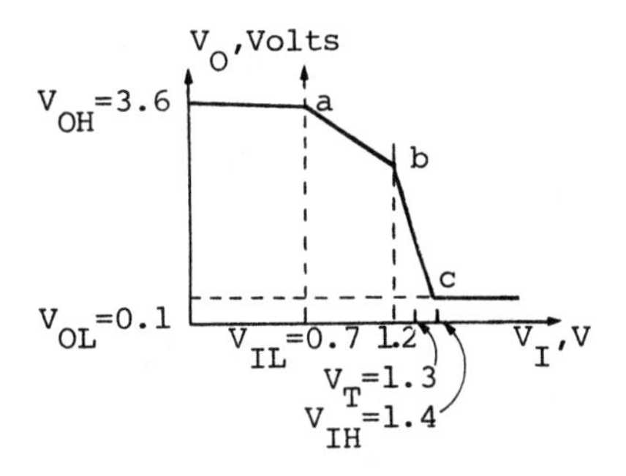

TTL Transfer curve

14.4.2 EMITTER-COUPLED LOGIC (ECL)

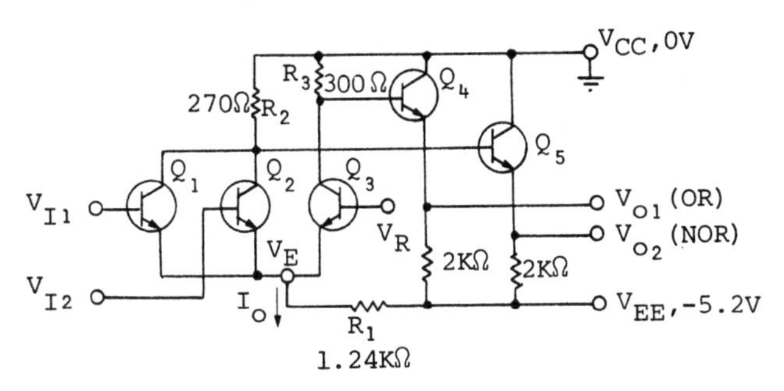

Basic ECL OR/NOR gate

CHAPTER 15

BOOLEAN ALGEBRA

15.1 LOGIC FUNCTIONS

15.1.1 NOT FUNCTION

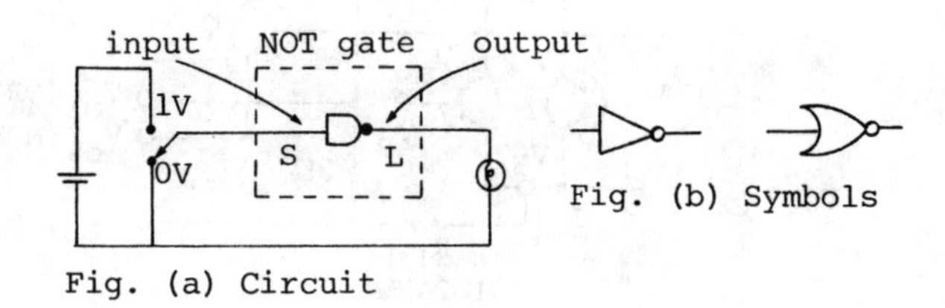

Fig. (a) Circuit

Fig. (b) Symbols

Logic equation and truth table:

$L = \overline{S}$

S	$L=\overline{S}$
1	0
0	1

15.1.2 AND FUNCTION

The AND functioning circuit:

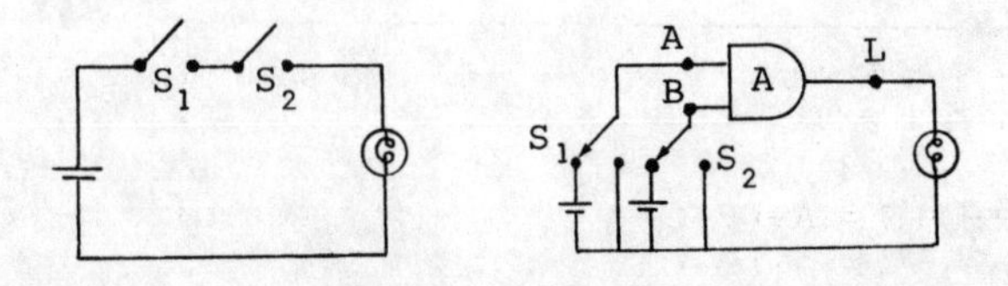

Logic equation and truth table:

$L = A * B$

A	B	L
0	0	0
0	1	0
1	0	0
1	1	1

15.1.3 OR FUNCTION

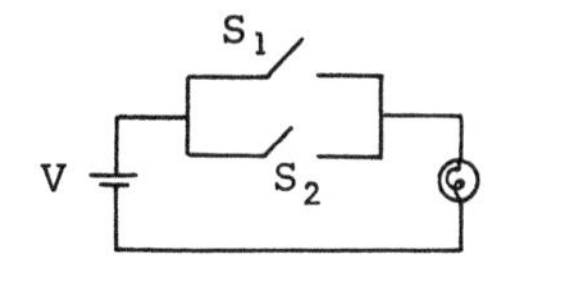

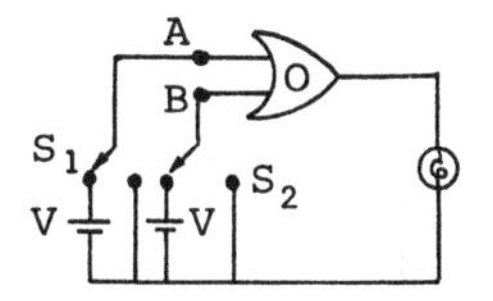

Logic equation and truth table:

$L = A + B$

A	B	L
0	0	0
0	1	1
1	0	1
1	1	1

15.2 BOOLEAN ALGEBRA

Boolean Theorems:

Theorem	Name
$A + B = B + A$ $A \cdot B = B \quad A$	Commutative law
$(A+B)+C = A+(B+C)$ $(A \cdot B) \cdot C = A \cdot (B \cdot C)$	Associative law

$A \cdot (B+C) = A \cdot B + A \cdot C$ Distributive law
$A + (B \cdot C) = (A+B) \cdot (A+C)$

$A + A = A$ Identity law
$A \cdot A = A$

$\overline{A} = \overline{A}$ Negation
$\overline{\overline{A}} = A$

$A + A \cdot B = A$ Redundancy
$A \cdot (A+B) = A$

$0 + A = A$
$1 \cdot A = A$
$1 + A = 1$
$0 \cdot A = 0$

$\overline{A} + A = 1$
$\overline{A} \cdot A = 0$

$A + \overline{A} \cdot B = A + B$
$A \cdot (\overline{A} + \overline{B}) = A \cdot \overline{B}$

$\overline{A+B} = \overline{A} \cdot \overline{B}$ De Morgan's laws
$\overline{A \cdot B} = \overline{A} + \overline{B}$

Proof of theorems:

Proof of theorem $(A+B) \cdot (A+C) = A + (B \cdot C)$

$$
\begin{aligned}
(A+B) \cdot (A+C) &= AA + AB + AC + BC \\
&= A + A(B+C) + BC \\
&= A(1+B+C) + BC \\
&= A + BC
\end{aligned}
$$

Proof of theorem $A + A \cdot B = A$

A	B	A • B	A+A • B
0	0	0	0
0	1	0	0
1	0	0	1
1	1	1	1

Proof of theorem $A + \bar{A} \cdot B = A + B$

A	B	A+B	$\bar{A}$	$\bar{A}B$	$A+\bar{A}B$
0	0	0	1	0	0
0	1	1	1	1	1
1	0	1	0	0	1
1	1	1	0	0	1

Proof of theorem $\overline{A+B} = \bar{A} \cdot \bar{B}$

A	B	A+B	$\overline{A+B}$	$\bar{A}$	$\bar{B}$	$\bar{A} \cdot \bar{B}$
0	0	0	1	1	1	1
0	1	1	0	1	0	0
1	0	1	0	0	1	0
1	1	1	0	0	0	0

Manipulations of logic equations:

Simplify $L = \bar{X}Y + XY + \bar{X}\,\bar{Y}$

$$= Y(X+\bar{X}) + \bar{X}\,\bar{Y}$$

$$= Y(1) + \bar{X}\bar{Y} \quad = Y + \bar{X}\,\bar{Y}$$

$$= Y + \bar{X}$$

If $L = \bar{X}Y + X\bar{Y}$, find $\bar{L}$.

$$\bar{L} = \overline{\bar{X}Y + X\bar{Y}}$$

$$= (\bar{X}Y)\ (X\bar{Y})$$

$$= (\bar{\bar{X}}+\bar{Y})(\bar{X}+\bar{\bar{Y}})$$

$$= (X+\bar{Y})(\bar{X}+Y)$$

$$= X\,\bar{X} + XY + \bar{Y}\,\bar{X} + \bar{Y}Y$$

$$= XY + \bar{Y}\,\bar{X}$$

15.3 THE NAND AND NOR FUNCTIONS

The NAND function:

Logic equation: $L = \overline{ABCD\ldots}$

The NAND operation is commutative, i.e., $L = \overline{ABC} = \overline{BAC} = \ldots$, but not associative.

Truth table:

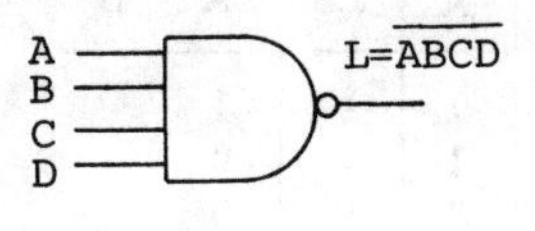

A	B	L	A·B
0	0	1	0
0	1	1	0
1	0	1	0
1	1	0	1

The NOR function:

The logic equation is $L = \overline{A+B}$

The NOR operation is commutative, i.e., $L = \overline{A+B+C\ldots} = \overline{B+A+C\ldots}$ but it is not associative.

Truth table:

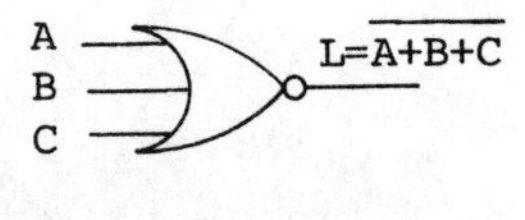

A	B	L	A+B
0	0	1	0
0	1	0	1
1	0	0	1
1	1	0	1

The exclusive OR function:

The logic equation is $L = A \oplus B = \bar{A}B + A\bar{B}$

The function is both commutative and associative.

In practice, these gates with more than two inputs are not available.

Truth table:

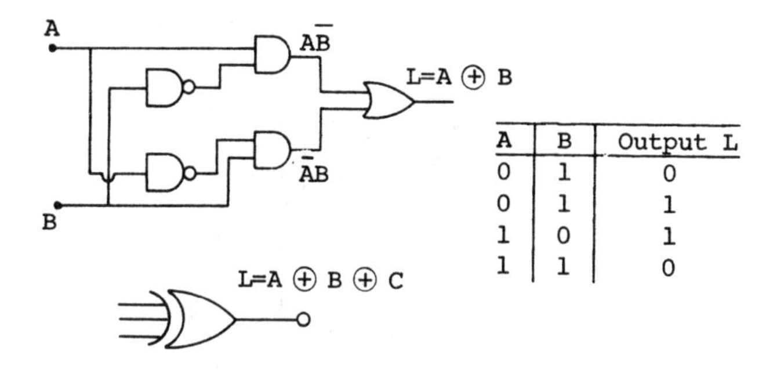

A	B	Output L
0	1	0
0	1	1
1	0	1
1	1	0

15.4 STANDARD FORMS FOR LOGIC FUNCTIONS

Sum of products (SP):

The logic function is written as a simple sum of terms, e.g.,

$$L = (\bar{W} + XY)(X + YZ)$$

$$= (\bar{W} + XY)X + (\bar{W} + XY)YZ$$

$$= \bar{W}X + XXY + \bar{W}YZ + XYYZ$$

$$= \bar{W}X + XY + \bar{W}YZ + XYZ \text{ (the desired form).}$$

To find the logic equation for L from the truth table:

Select the rows for which L = 1, these are called "Minterms."

Find the logic expression for these in the product form.

i.e., Row #3 (Minterm): $\bar{X}Y\bar{Z} = \bar{0}\ 1\ \bar{0} = 1$

Take the sum of all the minterms:

$$L = \overline{X}Y\overline{Z} + \overline{X}YZ + X\overline{Y}\overline{Z} + XY\overline{Z} + XYZ$$

Row #	X	Y	Z	L
1	0	0	0	0
2	0	0	1	0
3	0	1	0	1
4	0	1	1	1
5	1	0	0	1
6	1	0	1	0
7	1	1	0	1
8	1	1	1	1

Product of sums (PS):

This consists of a product of terms in which each term consist of a sum of all or part of the variables.

$$L = (\overline{W}+XY)(X+YZ) = (\overline{W}+X)(\overline{W}+Y)(X+Y)(X+Z)$$

To find a PS form from the truth table:

Pick-up rows for which L = 0, these are called the "Maxterms".

Express the logic expression for these in the sum form, i.e.,

For row #1: $X + Y + Z$

Take product of all these "Maxterms", i.e.,

$$L = (X+Y+Z)(X+Y+\overline{Z})(\overline{X}+Y+\overline{Z})$$

15.5 THE KARNAUGH MAP

It is a graphical technique for reducing logic equations to a minimal form.

Two variable Karnaugh map:

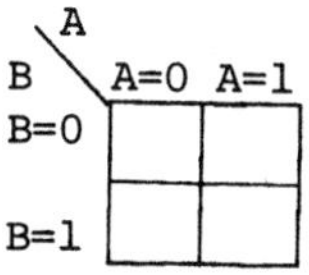

B \ A	A=0	A=1
B=0		
B=1		

Two-variable map

A	B	Term in logic equation = to 1
0	0	$\bar{A}\ \bar{B}$
0	1	$\bar{A}$ B
1	0	A $\bar{B}$
1	1	A B

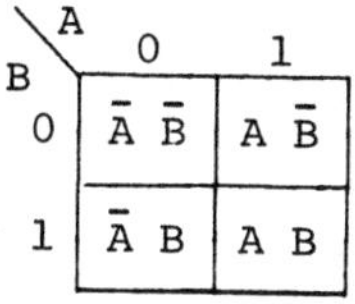

B \ A	0	1
0	$\bar{A}\ \bar{B}$	A $\bar{B}$
1	$\bar{A}$ B	A B

Correspondence with truth-table

Truth Table

A	B	L
0	0	0
0	1	1
1	0	1
1	1	1

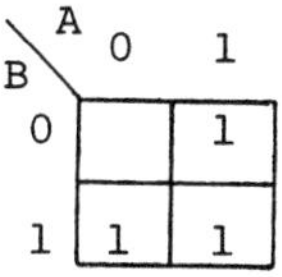

B \ A	0	1
0		1
1	1	1

K map for $L=\bar{A}B+A\bar{B}+AB$

An example

Set of rules for simplification:

A group of two adjacent cells combines to yield a single variable.

A single cell which can't be combined represents a two-variable term.

It is permissible for groups to overlap because of the fact that in boolean algebra, A+A = A.

Three-variables map:

AB / C	00	01	11	10
C→0	$\bar{A}\bar{B}\bar{C}$	$\bar{A}B\bar{C}$	$AB\bar{C}$	$A\bar{B}\bar{C}$
1	$\bar{A}\bar{B}C$	$\bar{A}BC$	ABC	$A\bar{B}C$

A primary map

The set of rules for simplification:

A group of 4 adjacent cells (in-line or square) combines to yield a single variable.

A group of two adjacent cells combines to yield a two-variable term.

A single cell which can't be combined represents a 3-variable term.

Use of these terms:

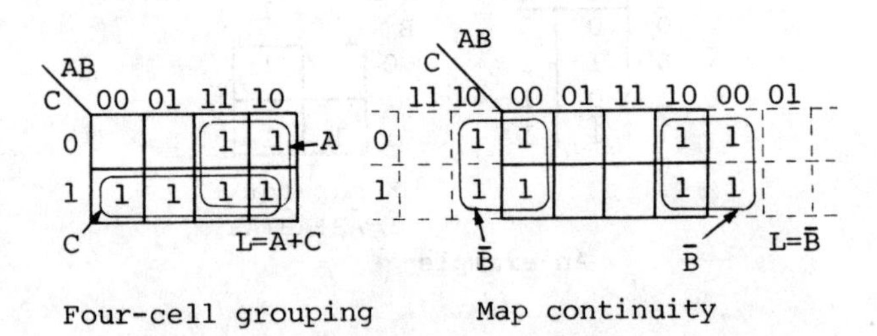

Four-cell grouping Map continuity

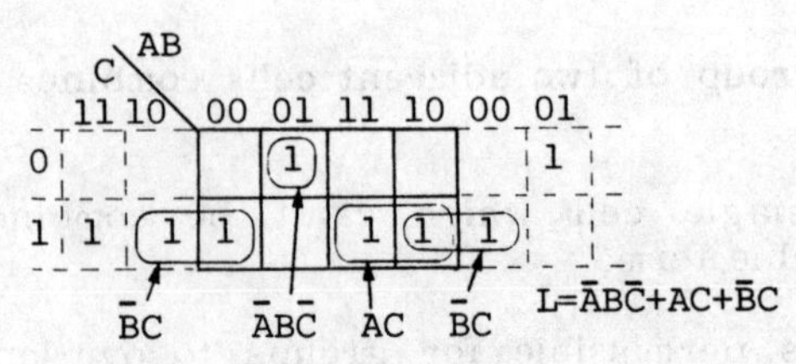

Two-cell groupings and single cell.

Four-variable map:

The rules are:

Eight adjacent cells yield a single variable. 4-adjacent cells yield a two-variable term. Two adjacent cells yield a 3-variable term. Individual cells represent 4-variable terms.

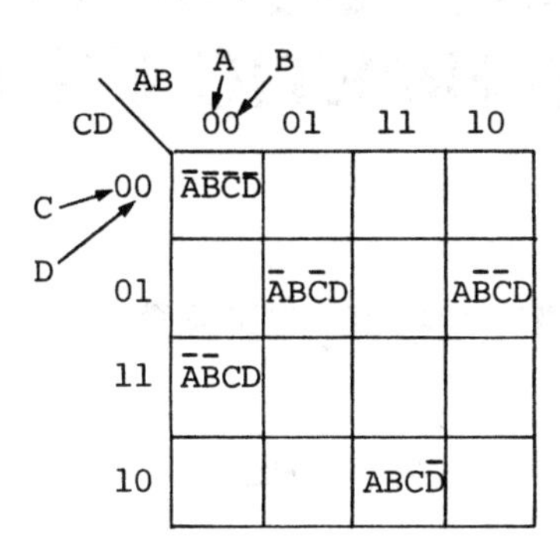

Primary map

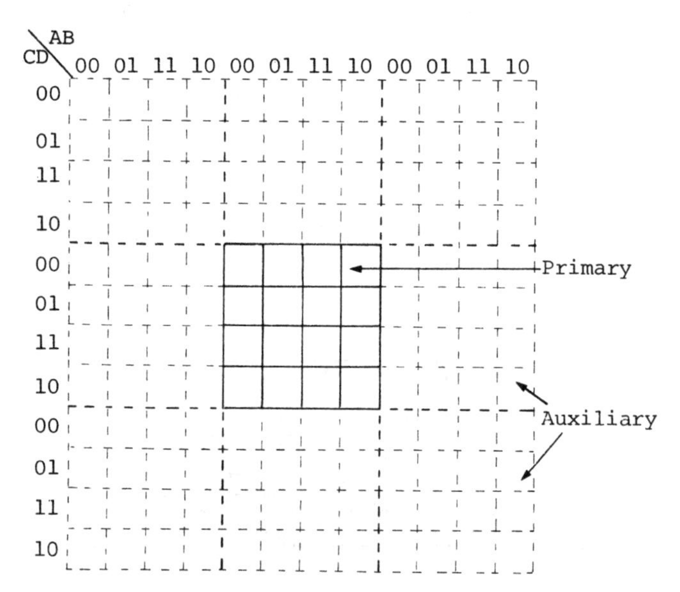

Continuity with auxiliary maps

CHAPTER 16

REGISTORS, COUNTERS AND ARITHMETIC UNITS

16.1 SHIFT REGISTERS

16.1.1 SERIAL-IN SHIFT REGISTERS

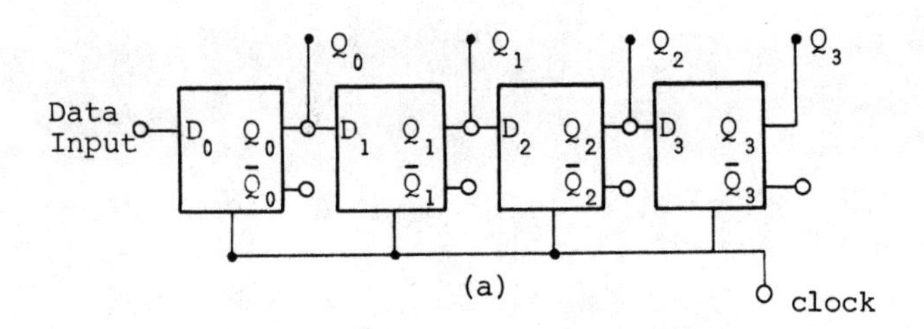

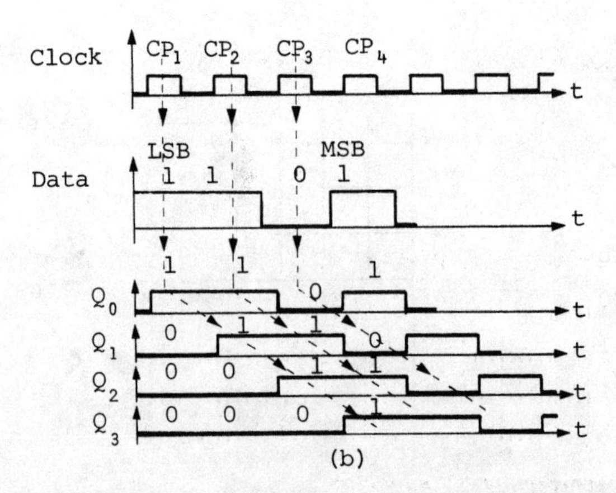

Four-bit shift registers: (a) logic diagram;
(b) waveforms

Data can be taken from this register in either serial or parallel form. For serial removal, it is necessary to apply four additional clock pulses; the data will then appear on Q_3 in serial form.

To read data in parallel form, it is only necessary to enter the data serially. Once the data are stored, each bit appears on a separate output line, Q_o to Q_3.

16.1.2 PARALLEL-IN SHIFT REGISTERS

The flip-flops have asynchronous preset and clear capability.

The unit has synchronous serial or asynchronous parallel-load capability and a clocked serial output.

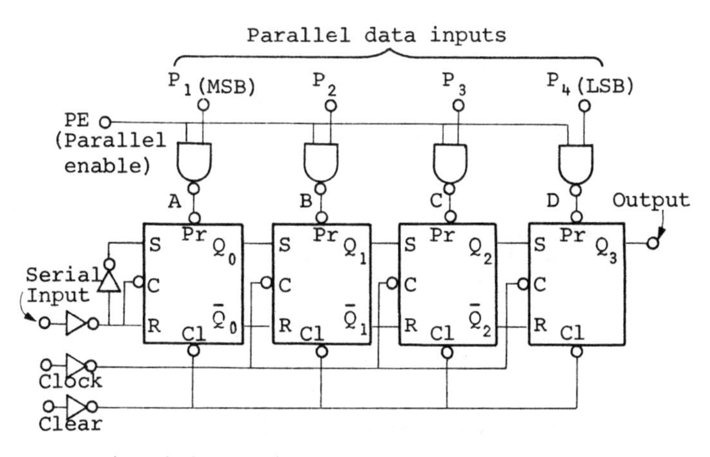

Simplified 54/7494 4-bit shift register.

16.1.3 UNIVERSAL SHIFT REGISTERS

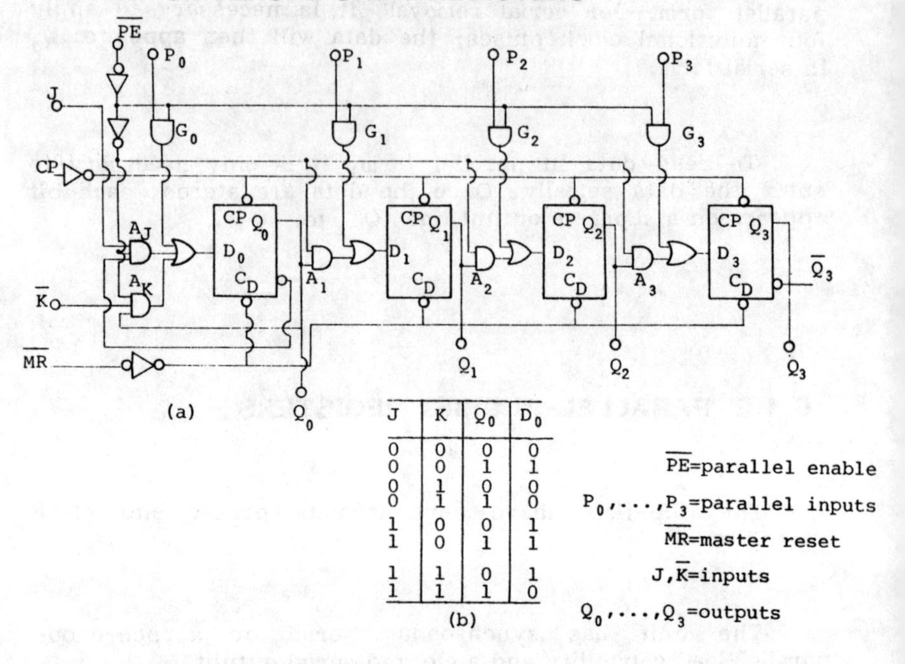

J	K	Q_0	D_0
0	0	0	0
0	0	1	1
0	1	0	0
0	1	1	0
1	0	0	1
1	0	1	1
1	1	0	1
1	1	1	0

This register contains four clocked master-slave flip-flops with D inputs.

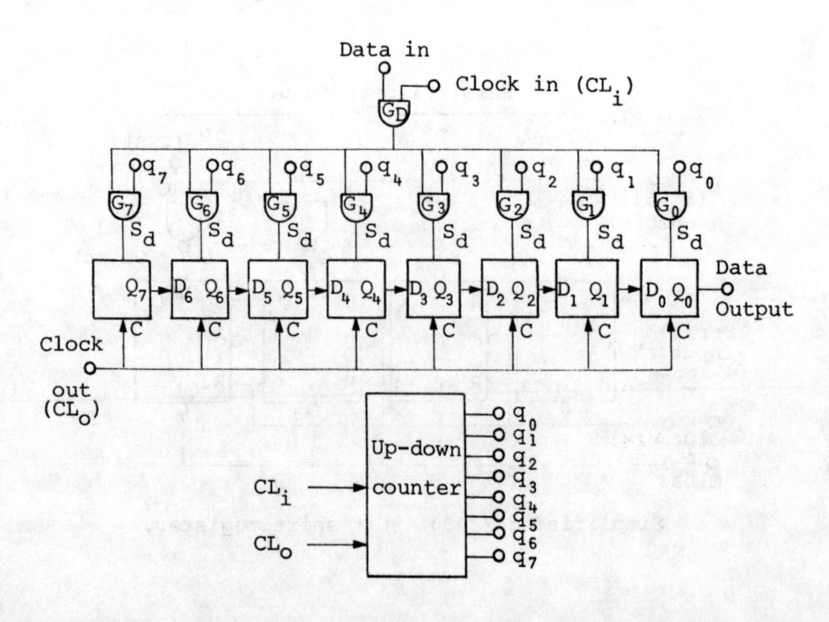

16.2 COUNTERS

16.2.1 THE RIPPLE COUNTER

This counter uses the maximum count capability of the three stages; hence, it is a mod-8 counter (the maximum modulus of an N-flipflop counter is 2^N).

Count $C = (Q_2 \times 2^2) + (Q_1 \times 2^1) + (Q_o \times 2^o)$

To obtain the decimal output, a binary-to-decimal decoder as shown below is used.

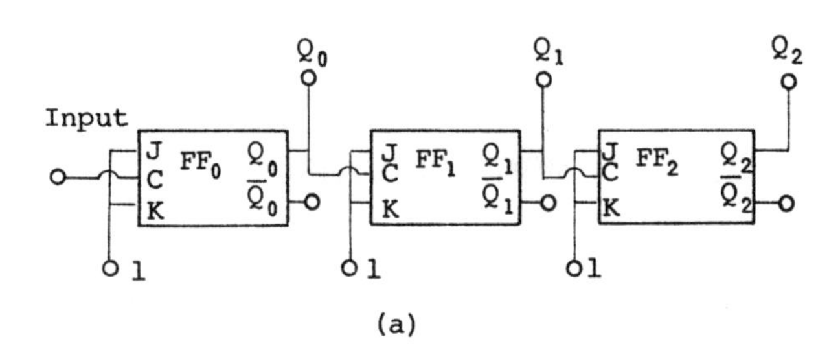

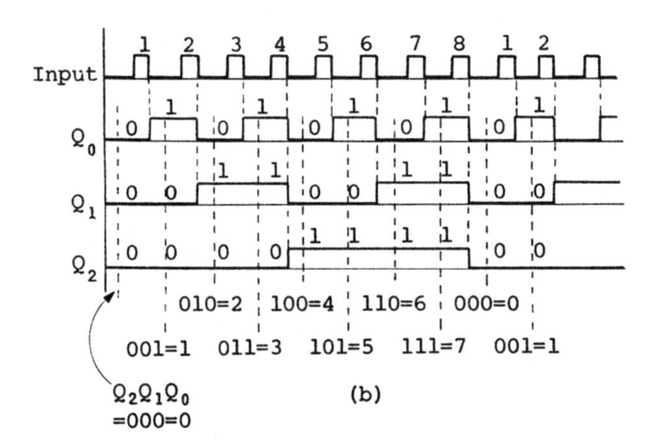

Mod-8 ripple counter: (a) logic diagram;
(b) waveforms.

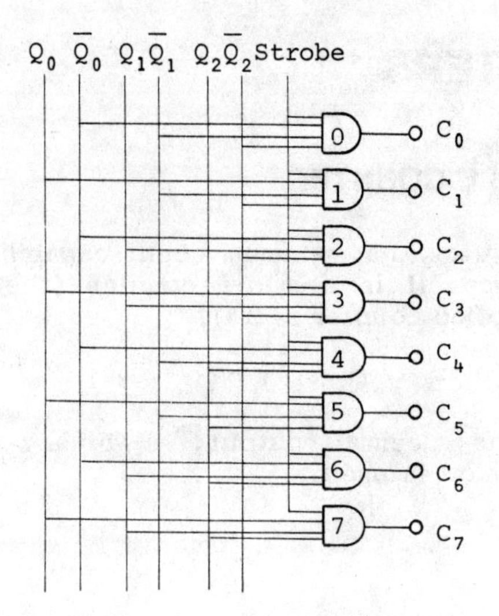

Binary-to-decimal decoder

16.2.2 THE SYNCHRONOUS COUNTER

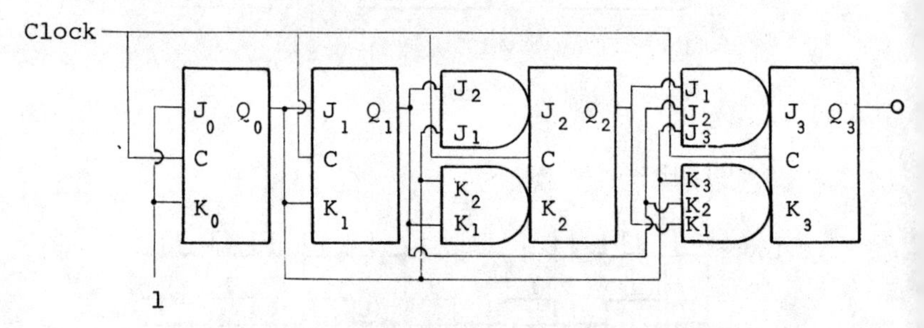

Circuit of synchronous parallel counter

The state table:

State table		
Counter state	Q_1 (2^1)	Q_0 (2^0)
0	0	0
1	0	1
2	1	0
0	0	0
1	0	1
2	1	0
0	0	0
⋮	⋮	⋮

FF_0 must change state (toggle) with each clock pulse. This is done by connecting J_0 and K_0 to a high level.

FF_1 must change state whenever $Q_0 = 1$. This is achieved by connecting J_1 and K_1 directly to Q_0.

FF_2 changes state only when $Q_0 = Q_1 = 1$. Thus Q_0 and Q_1 are connected through AND gates to J_2 and K_2.

FF_3 changes state only when $Q_0 = Q_1 = Q_2 = 1$. This requires 3-input AND gates connecting Q_0, Q_1 and Q_2 to J_3 and K_3.

Each following stage requires an additional input to the AND gate.

The propagation delay of an F-F increases with the load and limits the speed attainable with the counter.

$$f_{max} \leq \frac{1}{t_{pd}(FF_3) + t_{pd}(AND) + t_s}$$

16.3 ARITHMETIC CIRCUITS

16.3.1 ADDITION OF TWO BINARY DIGITS, THE HALF ADDER

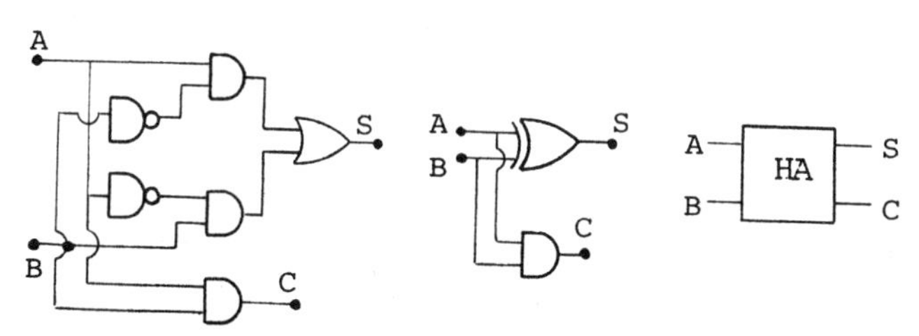

Circuit with AND, OR, & NOT gates

Truth Table:

A	B	S	C
0	0	0	0
0	1	1	0
1	0	1	0
1	1	0	1

The logic equations:

$$C = AB \text{ and } S = \bar{A}B + A\bar{B}$$

16.3.2 THE FULL ADDER

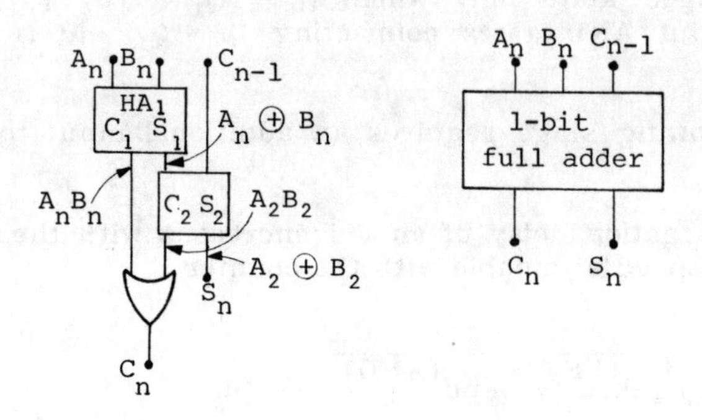

Truth table for adding A_n and B_n and a carry C_{n-1}.

A_n	B_n	C_{n-1}	S_n	C_n
0	0	0	0	0
0	1	0	1	0
1	0	0	1	0
1	1	0	0	1
0	0	1	1	0
0	1	1	0	1
1	0	1	0	1
1	1	1	1	1

Logic equations:

$$S_n = C_{n-1}(\bar{A}_n\bar{B}_n + A_nB_n) + \bar{C}_{n-1}(\bar{A}_nB_n + A_n\bar{B}_n)$$

$$= C_{n-1} \oplus (A_n \oplus B_n)$$

$$C_n = A_nB_n + C_{n-1}(A_n + B_n)$$

16.3.3 PARALLEL ADDITION

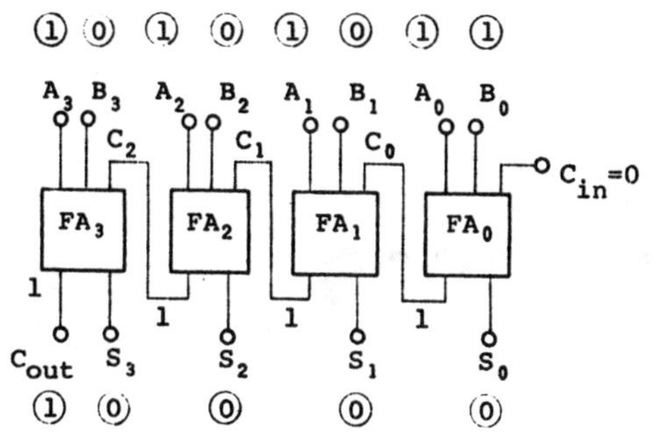

4-bit parallel adder.

If the propagation delay time $t_A = 2t_{pd}$ is the same for each adder, then the carry from FA_0 appears at FA_1 after time t_A, the carry from FA_1 appears at FA_2 after $2t_A$, and so on.

16.3.4 LOOK-AHEAD-CARRY ADDERS

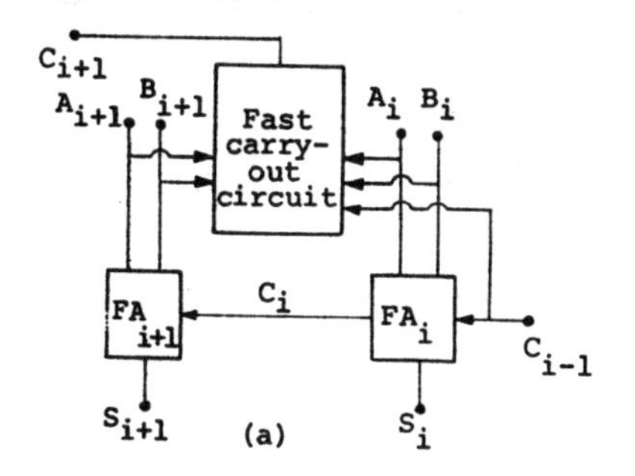

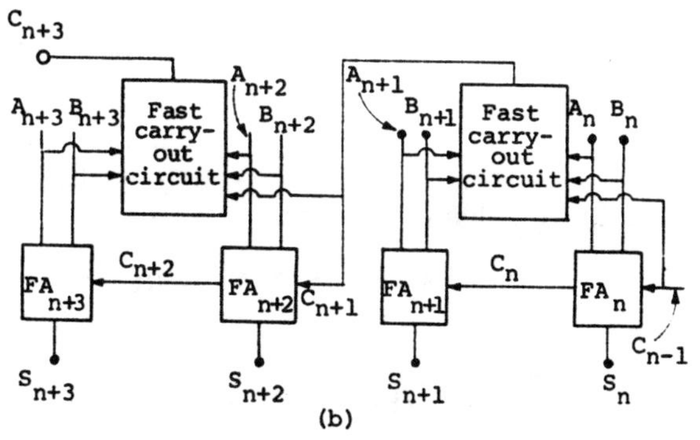

Carry-circuits: (a) 2-bit adder with a fast carry output; (b) 4-bit adder with fast carry output.

CHAPTER 17

OSCILLATORS

17.1 HARMONIC OSCILLATORS

A harmonic oscillator generates a sinusoidal output.

17.1.1 THE RC PHASE SHIFT OSCILLATOR

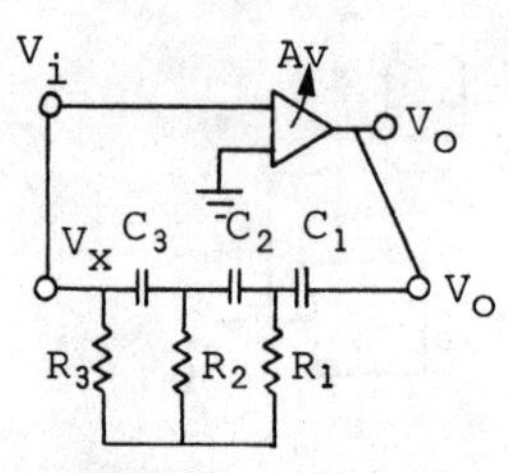

Amplifier with RC feedback network

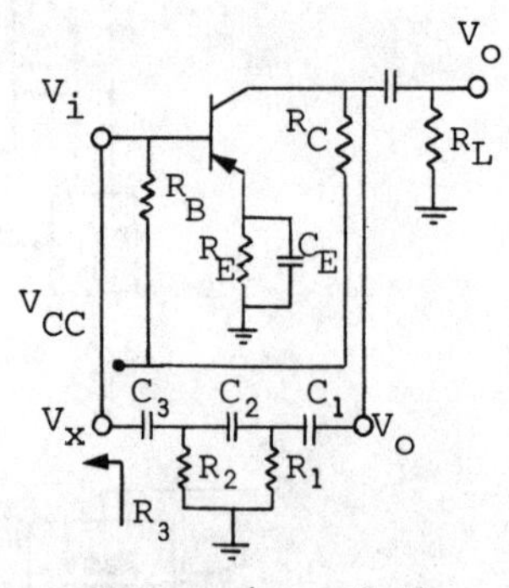

CE version of the circuit

For the feedback network:

$$R_3C_3 = R_2C_2 = R_1C_1 = RC$$

$$\frac{V_i}{V_o} = \frac{j\omega RC}{1+j\omega RC}$$

The overall transfer function $= \frac{V_x}{V_o} = \left(\frac{j\omega RC}{1+j\omega RC}\right)^3$

The circuit will oscillate at the frequency for which the phase shift from V_o to V_x is 180°.

$$f_o = 0.577/2\pi RC$$

$$\left|\frac{V_x}{V_o}\right| = \left(\frac{0.577}{1.58}\right)^3 = \frac{1}{8},$$

i.e., the voltage is attenuated by a factor of 8.

17.1.2 THE COLPITTS OSCILLATOR

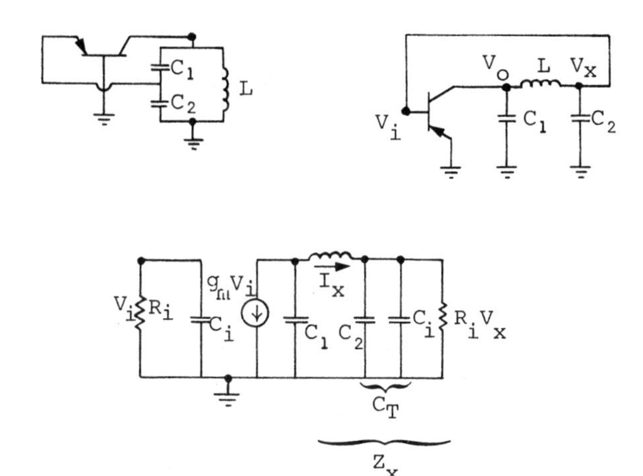

$$V_x = I_x \cdot Z_x, \quad I_x = -g_m \cdot V_i \cdot \frac{1/j\omega C_1}{(1/j\omega C_1)+j\omega L+Z_x}$$

$$V_x = V_i \text{ and } I_x - Z_x = \frac{I_x \; [1/j\omega C_1+j\omega L+Z_x]}{-g_m[1/j\omega C_1]}$$

$$Z_x = \frac{R_i[1/j\omega C_T]}{R_i+(1/j\omega C_T)}$$

$$1 + g_m R_i - \omega^2 LC_1 + j(\omega c_1 R_i + \omega C_T R_i - \omega^3 LC_1 C_T R_i) = 0$$

$$\omega^3 LC_1C_T \cdot R_i = \omega C_1R_i + \omega C_T \cdot R_i, \quad \omega^2 = \frac{C_1 + C_T}{L \cdot C_T \cdot C_1}$$

$$f_o = 1/2\sqrt{LC}, \quad \text{where } C = \frac{C_1 \cdot C_T}{C_1 + C_T}$$

Condition for sustained oscillations:

$g_m R_i = C_1/C_T$ (since $R_i \approx r_{b'e}$, $g_m R_i \approx g_m r_{b'e} = h_{fe}$),

$h_{fe} = C_1/C_T$

17.1.3 THE HARTLEY OSCILLATOR

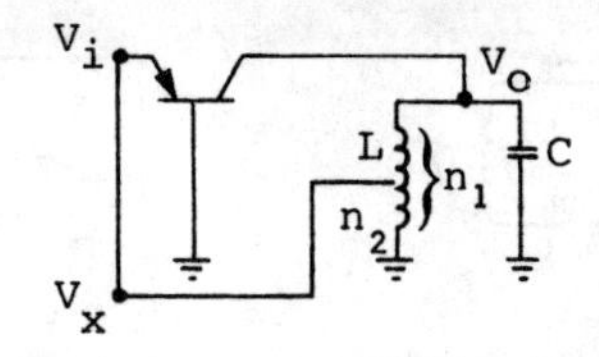

The feedback factor: $\beta_\nu = \frac{V_x}{V_o} = \frac{h_2}{h_1}$

The Hartley oscillator and its ac and output equivalent circuit:

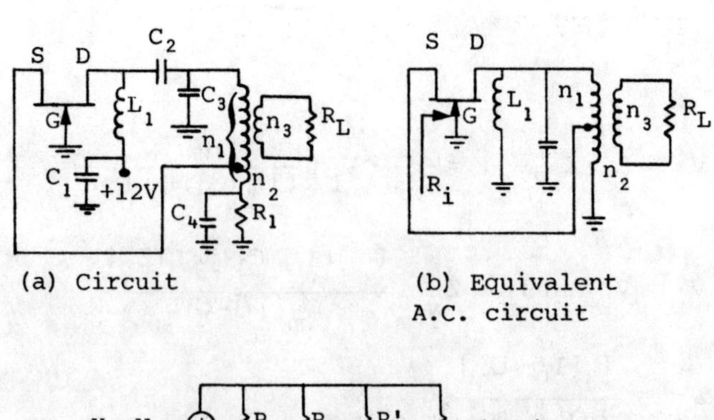

(a) Circuit

(b) Equivalent A.C. circuit

(c) Output equivalent circuit

The voltage amplification: $A_V \approx R'_L \, Y_{fs}$

The Barkhausen's criterion: $\dfrac{n_2 \cdot R'_L}{n_1} Y_{fs} = 1\angle 0°$

$$\frac{n_2}{n_1} = \frac{1}{R'_L \, Y_{fs}}$$

17.1.4 THE CLAPP OSCILLATOR

$$\left| j\omega_o L - \frac{j}{\omega_o C_3} \right| = \left| \frac{-j}{\omega_o C} \right|$$

$$\omega_o L - \frac{1}{\omega_o C_3} = 1/\,\omega_o C$$

$$f_o = 1/2\pi \sqrt{L \frac{C_3 C}{C_3+C}}$$

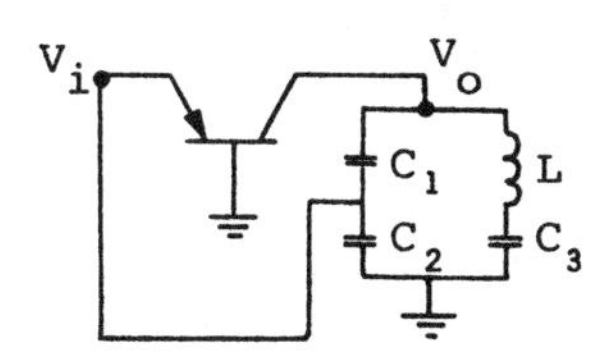

17.1.5 THE CRYSTAL OSCILLATOR

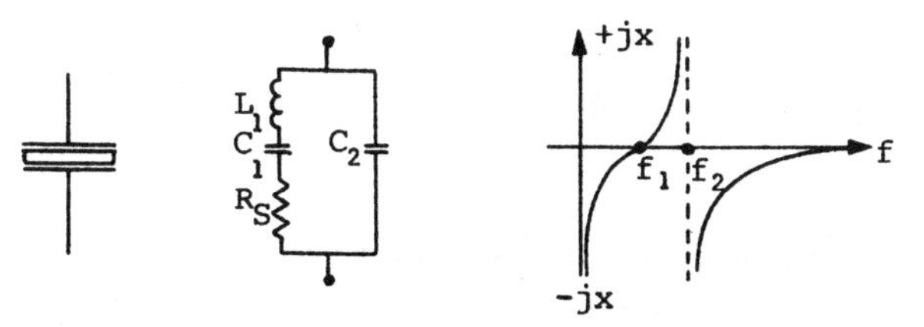

Resonant crystal, equivalent circuit and impedance characteristics

The series resonant frequency:

$f_1 = \dfrac{1}{2\pi\sqrt{L_1 C_1}}$, f_2 = The parallel resonant

$$\text{frequency} = \frac{1}{2\pi\sqrt{L_1C_1C_2/(C_1+C_2)}} = f_1\sqrt{1+\frac{C_1}{C_2}}$$

The Clapp and Colpitts crystal oscillators:

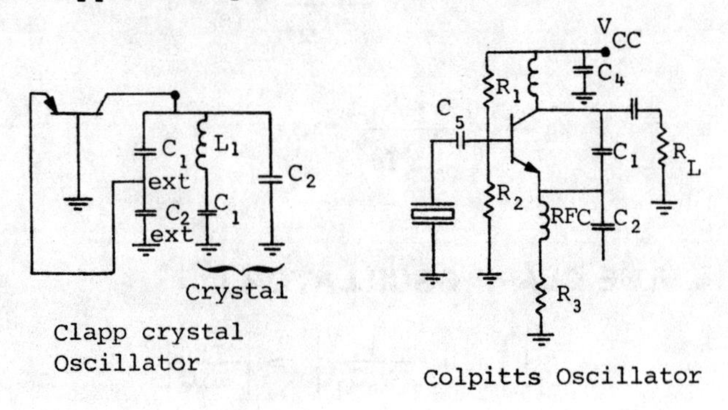

17.1.6 TUNNEL DIODE OSCILLATORS

These oscillators exhibit negative resistance when suitably biased.

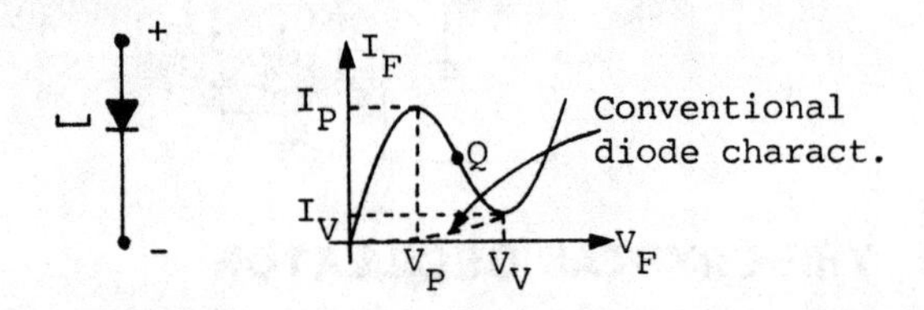

Dynamic resistance: $r = \Delta V_F / \Delta I_F$.

Tunnel diode dc biasing circuit:

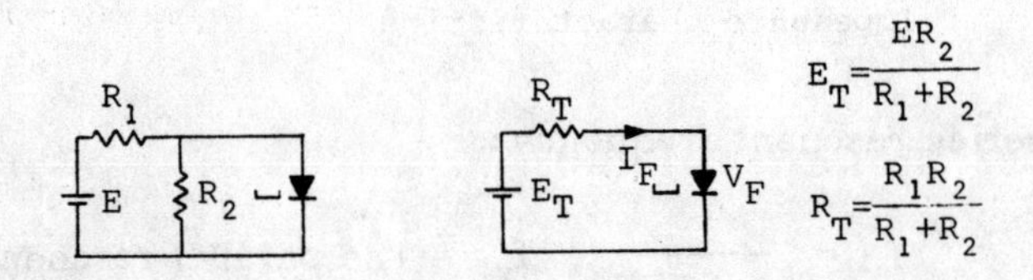

$$E_T = \frac{ER_2}{R_1+R_2}$$

$$R_T = \frac{R_1R_2}{R_1+R_2}$$

Tunnel diode oscillator:

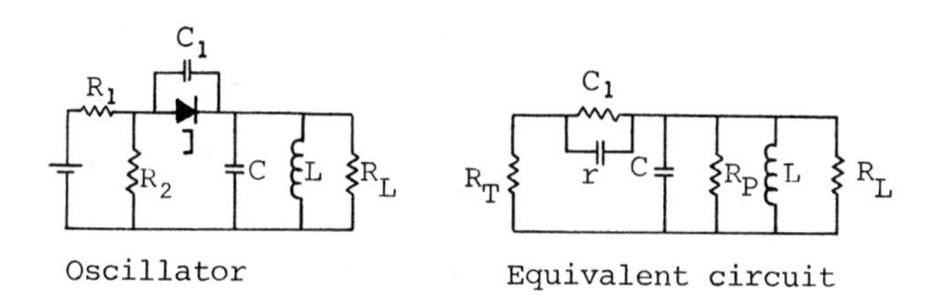

AC equivalent circuit for deriving criteria for oscillations:

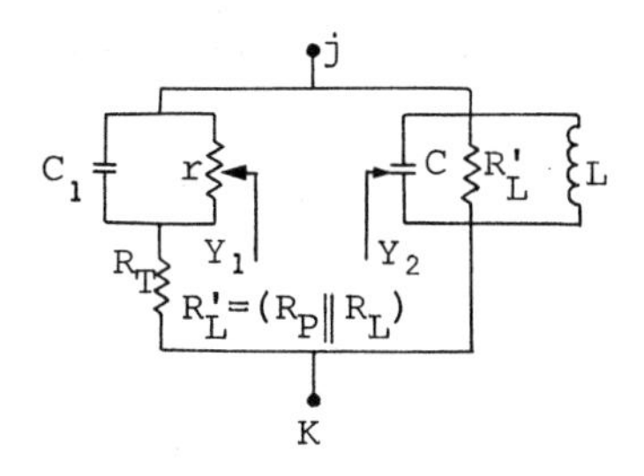

$$Y_{jk} = Y_1 + Y_2, \quad Y_1 = 1/Z_1$$

$$Z_1 = R_T + \left(r \parallel \frac{1}{j\omega c_1} \right)$$

$$R_T = \frac{-r}{1+\omega^2 r^2 C_1^{\,2}} \quad \text{(The condition for oscillations at } \omega \text{)}$$

$$f_{max}^{\;2} = \frac{-(r+R_T)}{r^2 R_T C_1^{\,2} (2\pi)^2} = \text{The highest possible frequency.}$$

$$f_o \text{ (The actual frequency of oscillations)} = \frac{1}{2\pi\sqrt{LC}} \cdot \frac{1}{\sqrt{1 + \dfrac{r \cdot C_1}{C(r+R_T)}}}$$

17.2 RELAXATION OSCILLATORS

The unijunction transistor (UJT) oscillator:

Useful relationships:

$R_{BB} = R_{B1} + R_{B2}$ (R_{BB} is the interbase resistance)

n = The standoff ratio = $R_{B1} / R_{B1} + R_{B2}$

Equivalent circuit and characteristics:

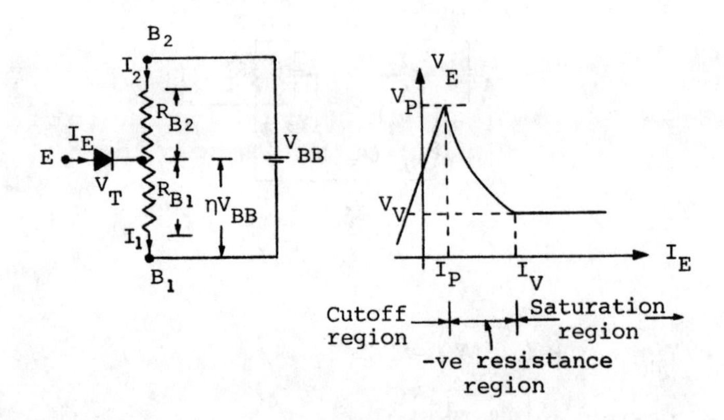

UJT oscillator:

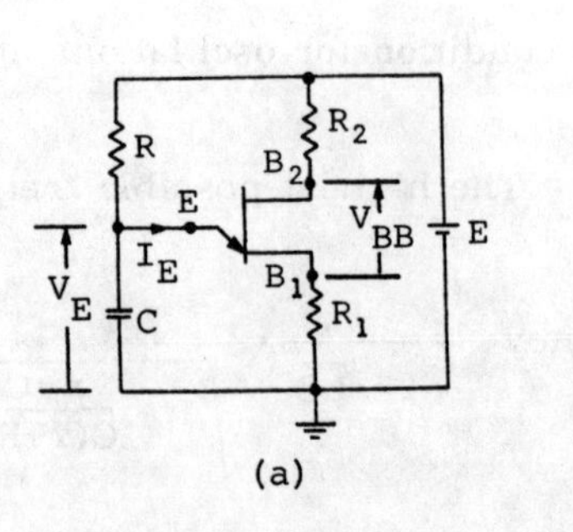

(a)

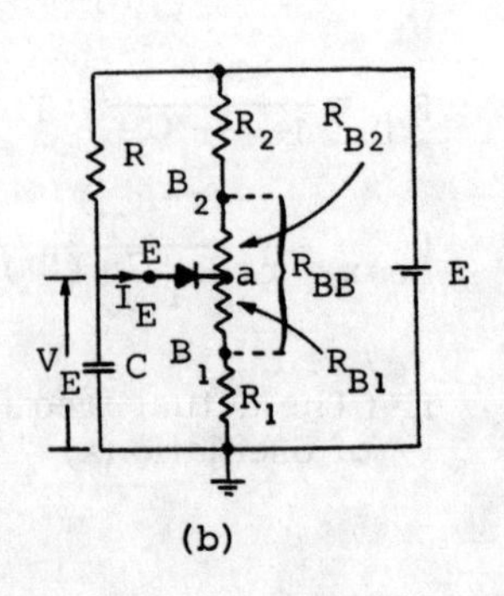

(b)

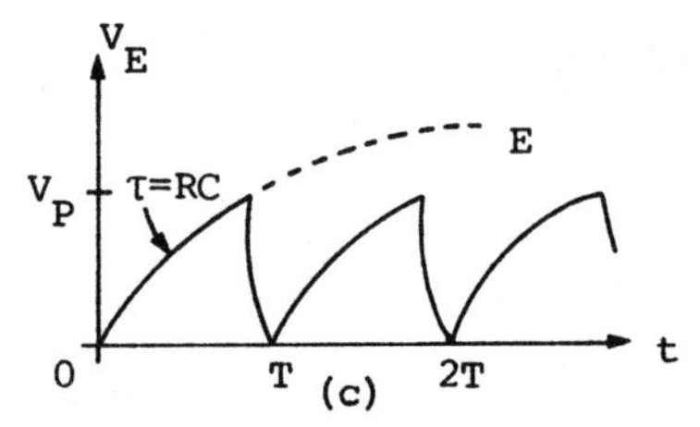

(a) Circuit of UJT oscillator; (b) AC equivalent circuit for (a); (c) emitter voltage waveform for the circuit of (a).

$$T = (R \cdot C)\ln\left|\frac{1}{1-n}\right|, \quad K = \ln\left|\frac{1}{1-n}\right|$$

f_o = The frequency of oscillations

$= 1/RCK$

Plot of K as a function of n:

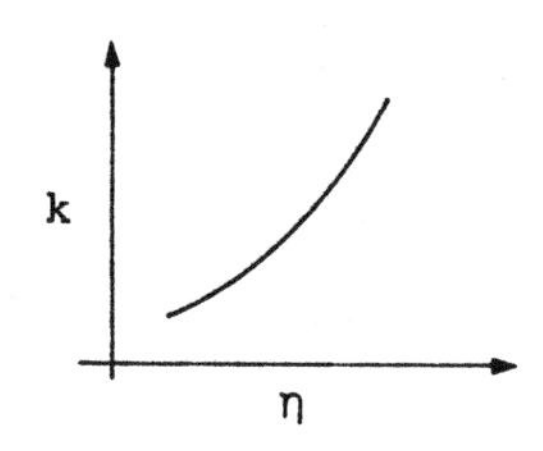

CHAPTER 18

RADIO-FREQUENCY CIRCUITS

18.1 NON-LINEAR CIRCUITS

A non-linear circuit is defined as any circuit that produces frequencies not present in the input signal.

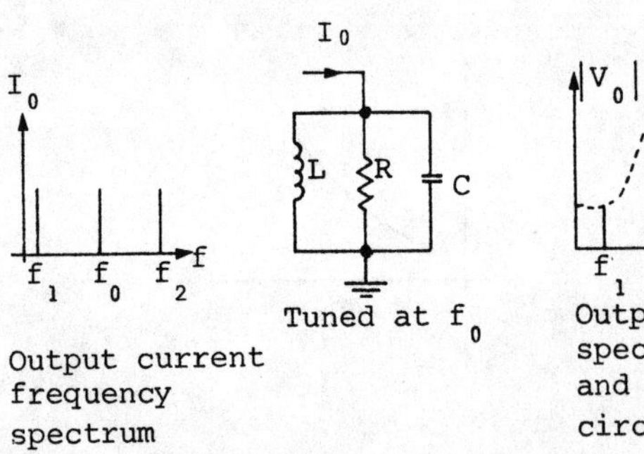

Output current frequency spectrum

Tuned at f_0

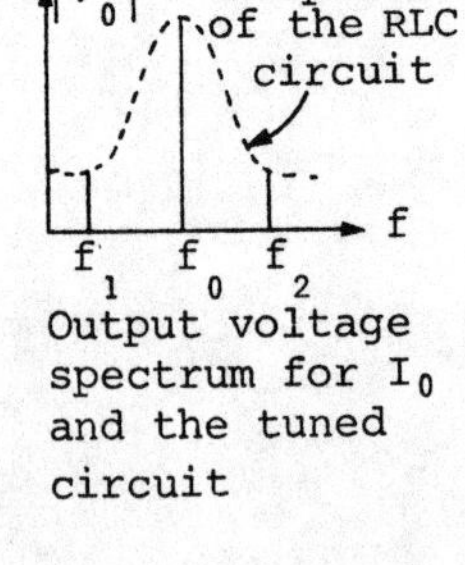

Output voltage spectrum for I_0 and the tuned circuit

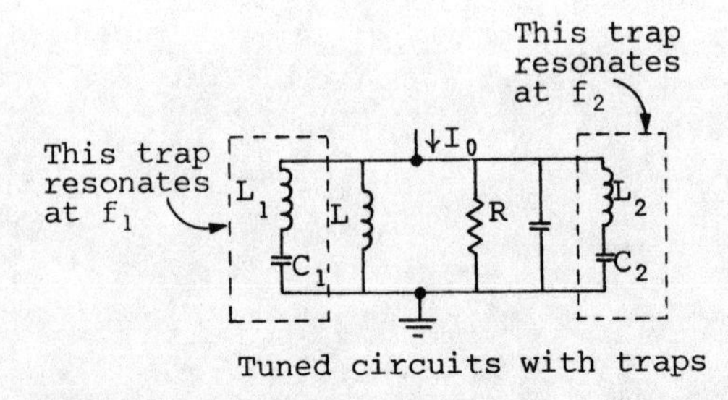

Tuned circuits with traps

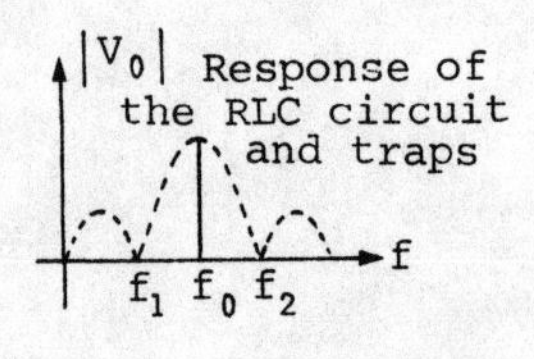

Class 'C' amplification:

input signal

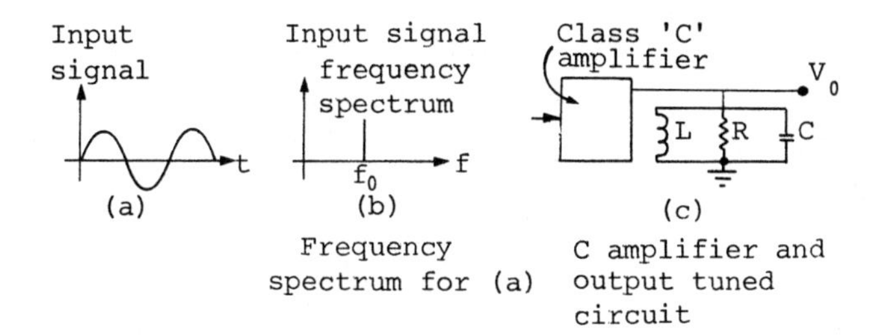

Waveforms and frequency spectra for a class 'C' circuit:

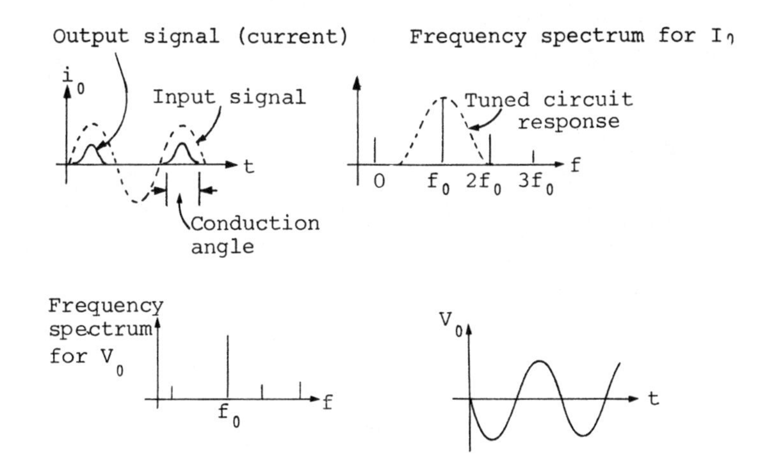

Frequency multiplication:

Diagram and signals for a frequency multiplier:

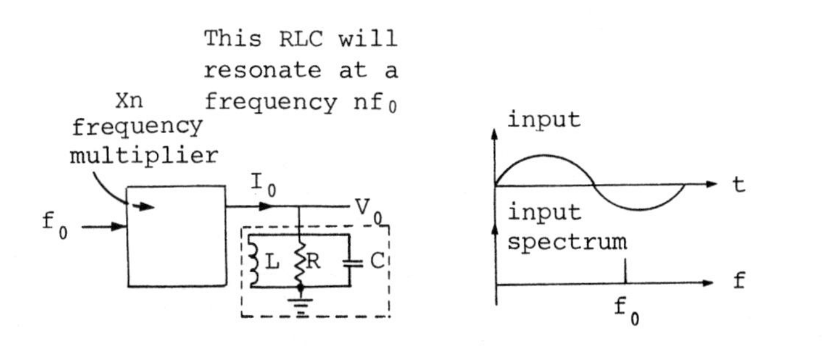

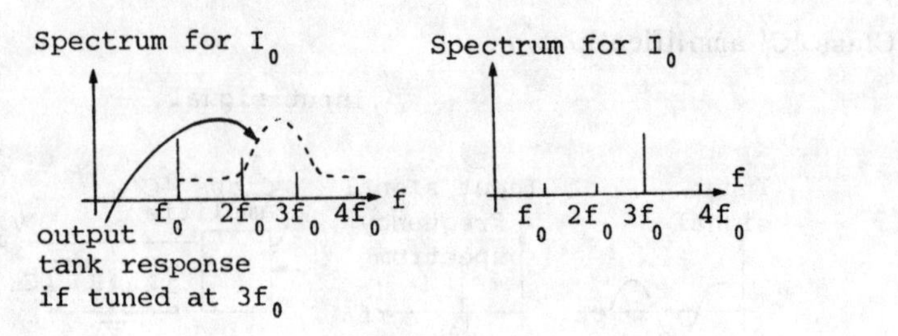

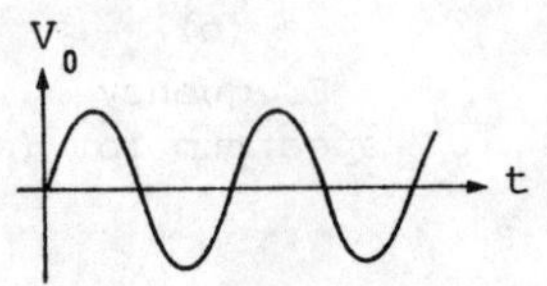

Mixing

Two signals are combined to yield an output signal with a frequency content which includes both the sum and difference of the two input frequencies.

V_o: Amplitude-modulated carrier

Carrier (frequency: f_0)

Active Device

I_0 V_0

Modulating signal (frequency: f_s)

C R L

A B A B

t

0

Block diagram of amplitude modulation system

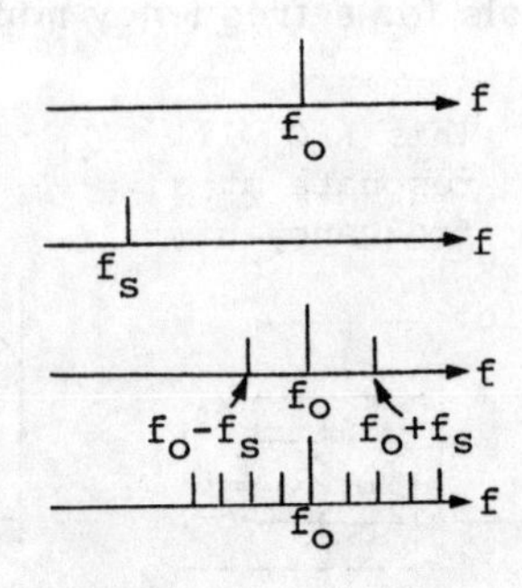

18.2 SMALL-SIGNAL RF AMPLIFIERS

Tuned amplifiers are mainly characterized by 1) center frequency f_o; 2) a bandwidth (B of 3dB 3) power gain at f_o: G_o and a noise figure at f_o, NF.

Basic circuits:

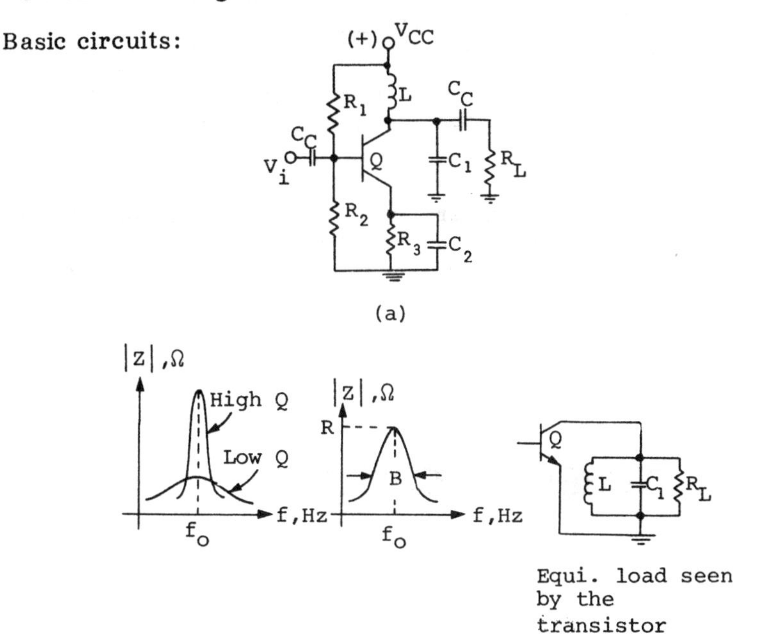

Bandwidth is a function of the circuit Q. A high Q means a narrow bandwidth, while a lower Q results in a wider bandwidth.

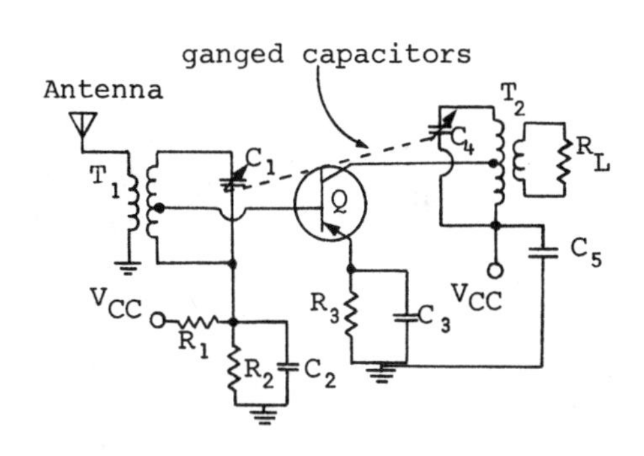

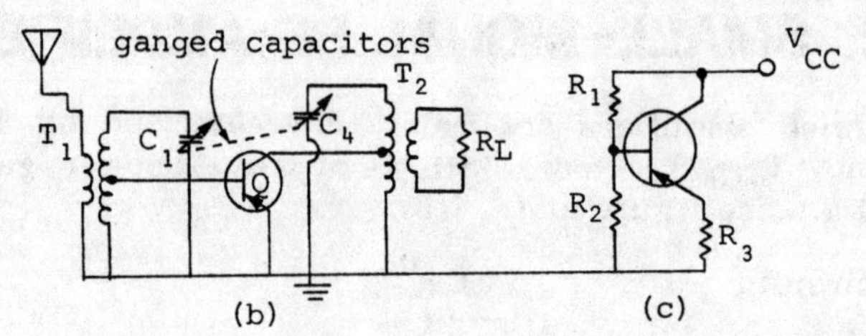

(a) RF amplifier with tuned input and output;
(b) AC circuit for (a); (c) DC circuit for (a)

Both the transformers are wound around an air core, that is, a non-magnetic core. The air core is used at high frequencies because the inductance required is small and the core losses are small, yielding a high Q.

18.3 TUNED CIRCUITS

$$Q = \frac{X_s}{R_s}, \quad Z_s = R_s \pm jX_s, \quad Z_p = (R_p)(\pm jX_p)/R_p \pm jX_p.$$

$$Z_s = Z_p \rightarrow R_s = \frac{R_p X_p^2}{R_p^2 + X_p^2}, \quad X_s = \frac{X_p R_p^2}{R_p^2 + X_p^2}$$

$$Q = \frac{R_p}{X_p}$$

18.4 CIRCUITS EMPLOYING BIPOLAR TRANSISTORS

Figure:

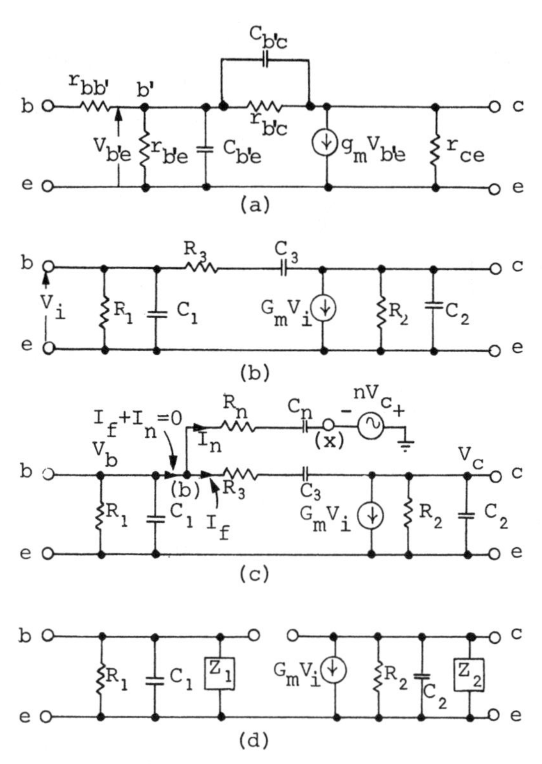

(a) Transistor hybrid-pi model; (b) transformed version of the circuit in (a); (c) unilateralized circuit; (d) unilateralized circuit including loading on input and output due to R_3, C_3, R_n, and C_n.

The circuit of Fig. (a) is transformed to (b). The parameters are:

$$R_1 = r_{bb'} + r_{b'e}\left[\frac{r_{b'e} + r_{bb'}}{r_{b'e} + r_{bb'} + \omega^2 (C_{b'e} + C_{b'c})^2 (r_{bb'} r_{b'e}{}^2)}\right]$$

$$C_1 = C_{b'e} + C_{b'c} / \left(1 + \frac{r_{bb'}}{r_{b'e}}\right)^2 + \omega^2 r_{bb'}^2 (C_{b'e} + C_{b'c})^2$$

$$R_2 = \left[\frac{1}{r_{ce}} + \frac{1}{r_{b'c}} + g_m \frac{\frac{1}{r_{b'c}} \left(\frac{1}{r_{bb'}} + \frac{1}{r_{b'e}}\right) + \omega^2 C_{b'e} C_{b'c}}{\left(\frac{1}{r_{bb'}} + \frac{1}{r_{b'e}}\right)^2 + \omega^2 C_{b'e}^2} \right]^{-1}$$

$$C_2 = C_{b'c} \left[1 + \frac{g_m \left(\frac{1}{r_{bb'}} + \frac{1}{r_{b'e}}\right)}{\left(\frac{1}{r_{bb'}} + \frac{1}{r_{b'c}}\right)^2 + \omega^2 C_{b'e}^2} \right]$$

$$R_3 = r_{bb'} \left(1 + \frac{C_{b'e}}{C_{b'c}}\right) + \frac{1 + r_{bb'}/r_{b'e}}{r_{b'c} \cdot \omega^2 \cdot C_{b'c}^2}$$

$$C_3 = C_{b'c} / \left[1 + \frac{r_{bb'}}{r_{b'e}} - \frac{r_{bb'}}{r_{b'c}} \frac{C_{b'e}}{C_{b'c}} \right]$$

$$G_m = g_m / \sqrt{\left(1 + \frac{r_{bb'}}{r_{b'e}}\right)^2 + [r_{bb'} \cdot \omega(C_{b'e} + C_{b'c})]^2}$$

Ideally, the amplifier should be unilateral, i.e., it should be possible for a signal to be transmitted from the input to the output, but not vice versa.

A method for unilateralizing a transistor:

$I_f = -I_n$ (so that there is no net current leaving point b)

$$\frac{V_b - V_c}{R_3 + (1/j\omega C_3)} = - \frac{V_b - (-n \cdot v_c)}{R_n + (1/j\omega C_n)}$$

$$\frac{-V_c}{R_3 + (1/j\omega C_3)} \approx \frac{n\ v_c}{R_n + (1/j\omega C_n)}$$

$$R_n = n \cdot R_3 \quad \text{and} \quad C_n = \frac{C_3}{n}, \quad n = \frac{V_x}{V_c}$$

Once the circuit is unilateral, the input and output may be treated independently (Fig. d).

18.5 ANALYSIS USING ADMITTANCE PARAMETERS

The performance of an active device is predicted using Y-parameters, because these can be obtained at any specific frequency.

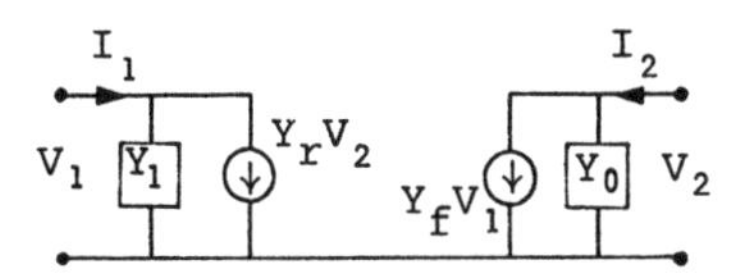

Using Y-parameters, because these can be obtained at any specific frequency.

$$Y_i = \frac{I_1}{V_1}\bigg|_{V_2=0}, \quad Y_f = \frac{I_2}{V_1}\bigg|_{V_1=0}$$

$$Y_r = \frac{I_1}{V_2}\bigg|_{V_1=0}, \quad Y_o = \frac{I_2}{V_2}\bigg|_{V_1=0}$$

A Y-parameter model with source and load added:

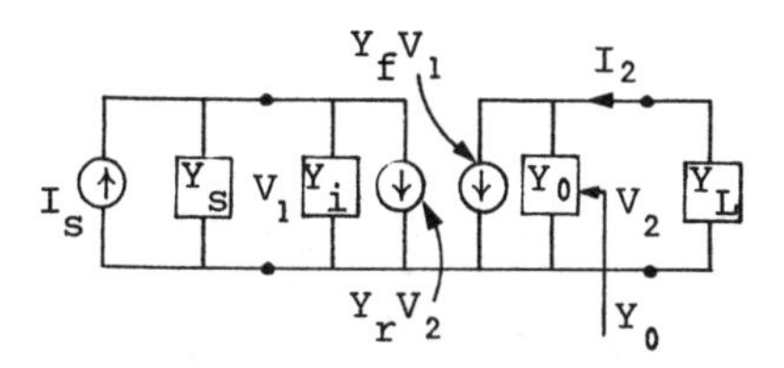

$$V_2 = -Y_f \cdot V_1 / Y_o + Y_L, \quad I_1 = (V_i \cdot Y_i) + \frac{-Y_f \cdot V_1}{Y_o + Y_L} Y_r$$

$$Y_i = \frac{I_1}{V_1} = Y_i - \left(\frac{Y_f \cdot Y_r}{Y_o + Y_L} \right)$$

$$V_1 = -V_2 Y_r / Y_i + Y_s, \quad I_2 = \frac{-Y_f V_2 Y_r}{Y_i + Y_s} + (Y_o V_2)$$

$$Y_o = Y_o - (Y_f Y_r / Y_i + Y_s)$$

CHAPTER 19

FLIP-FLOPS

19.1 TYPES OF FLIP-FLOPS

19.1.1 THE BASIC FLIP-FLOP

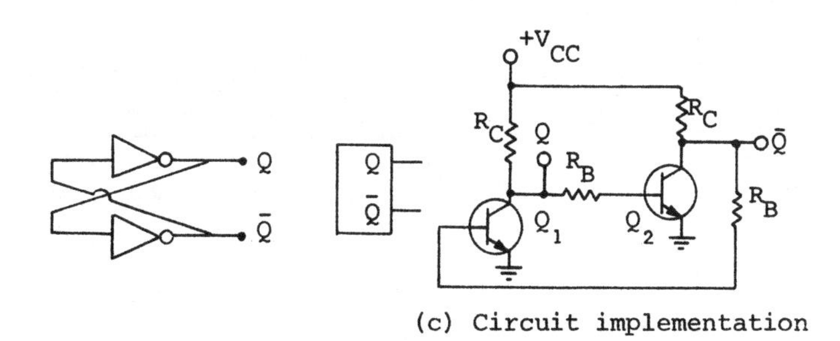

(c) Circuit implementation

The basic flip-flop simply consists of two RTL inverters.

19.1.2 R-S FLIP-FLOP

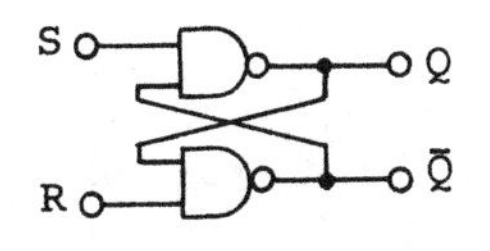

R	S	Q
1	1	unchanged
1	0	1
0	1	0
0	0	undefined

19.1.3 SYNCHRONOUS R-S FLIP-FLOP (CLOCKED R-S FLIP-FLOP)

The output state is set in synchronization with clock pulses.

NAND-gate implementation:

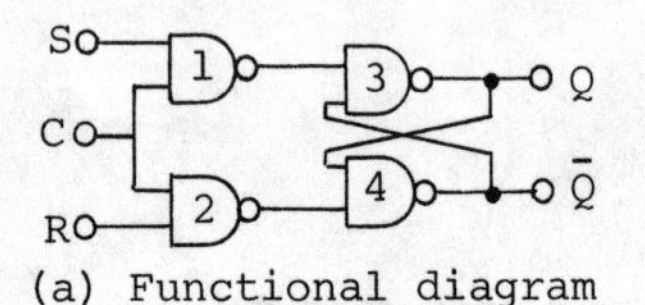

(a) Functional diagram

R_n	S_n	Q_{n+1}
0	0	Q_n
0	1	1
1	0	0
1	1	ND

(b) Truth table

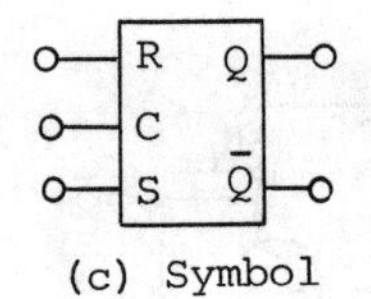

(c) Symbol

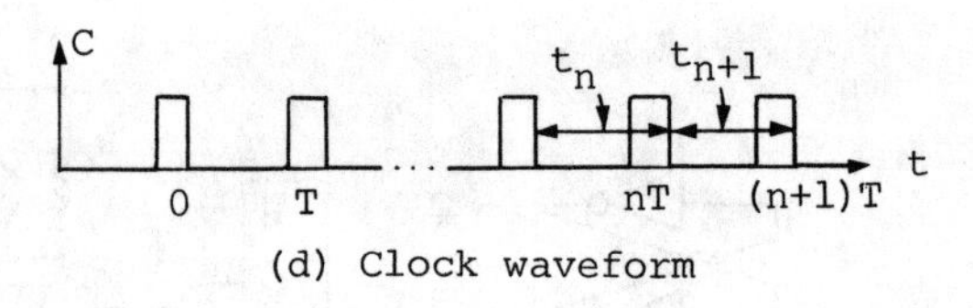

(d) Clock waveform

Gates 3 and 4 form an R-S latch with steering gates 1 and 2, used to input data to the device.

The clock is normally low, the outputs of gates 1 and 2 are then normally high and the state of the flip-flop cannot change. When C is high, the R and S inputs are steered to gates 3 and 4 and the flip-flop responds according to the truth table.

If R and S are low, the state of the flip-flop does not change, while Q in t_{n+1} is the same as it was in t_n, i.e., $Q_{n+1} = Q_n$.

19.1.4 PRESET AND CLEAR

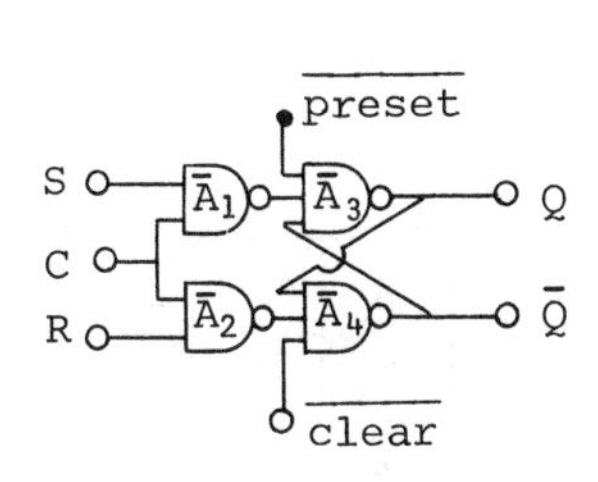

Synchronous		
R_n	S_n	Q_{n+1}
0	0	Q_n
0	1	1
1	0	0
1	1	ND
Asynchronous		
$\overline{clear}$	$\overline{preset}$	Q
0	1	0
1	0	1

(b) Truth table

General R-S Flip-Flop Signal Polarities
If S is high, Q=1
If R is high, Q=0
If $\overline{preset}$ is low, Q=1
If $\overline{clear}$ is low, Q=0

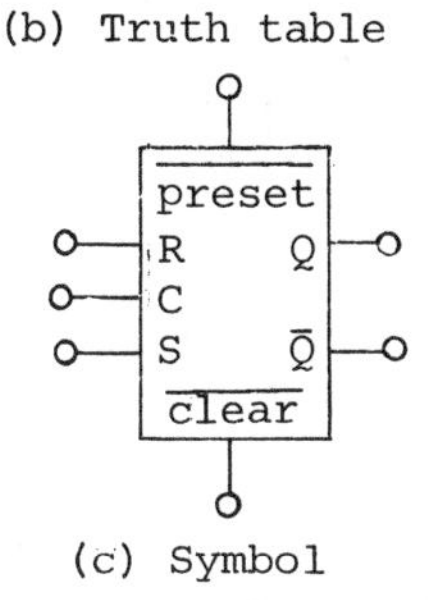

(c) Symbol

The two dc control lines are normally high.

The functional diagram: The two dc control lines do not require the clockpulse and, in fact, override the clocked inputs.

The asynchronous or dc truth table holds for the dc $\overline{clear}$ and $\overline{preset}$ inputs.

19.1.5 D-TYPE FLIP-FLOP

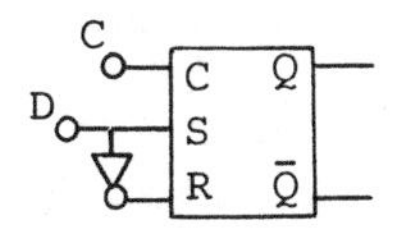

D_n	Q_{n+1}
1	1
0	0

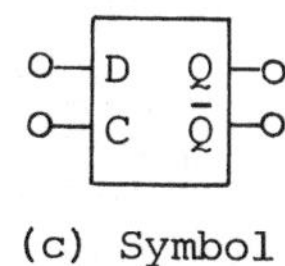

(c) Symbol

With the D input connected to the S input and an inverter between R and S, the undefined output state that occurred in the R-S flip-flop is not possible here.

19.1.6 J-K FLIP-FLOP

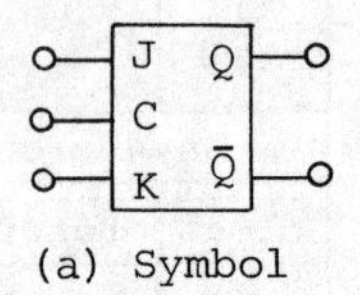

(a) Symbol

J_n	K_n	Q_{n+1}
0	0	Q_n
0	1	0
1	0	1
1	1	$\bar{Q}_n$

(b) Truth table

Q_n	Q_{n+1}	J_n	K_n
0	0	0	X
0	1	1	X
1	0	X	1
1	1	X	0

(c) Excitation table, X = don't care

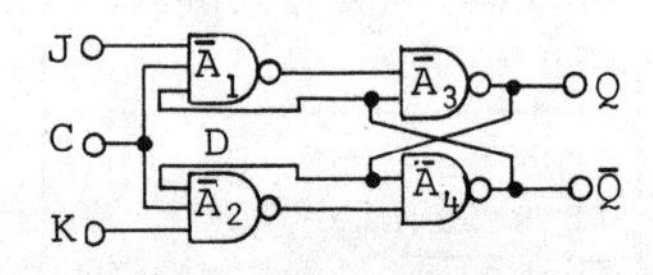

The J and K inputs listed in the truth table are the inputs present during time t_n. The output state is the flip-flop's state during t_{n+1} after a clock pulse at nT.

Excitation table: If the state Q_n of the flip-flop before clocking is known and the desired state Q_{n+1} after clocking is known, the necessary J and K inputs can be read from this table.

19.1.7 T-TYPE FLIP-FLOP

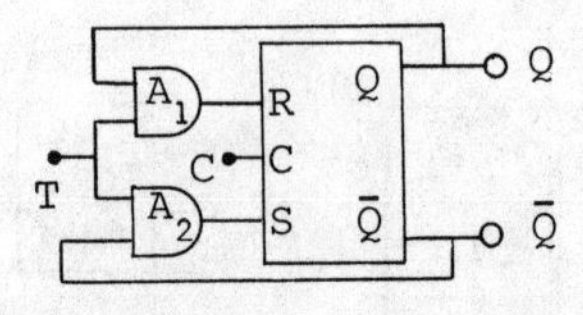

Functional Diagram

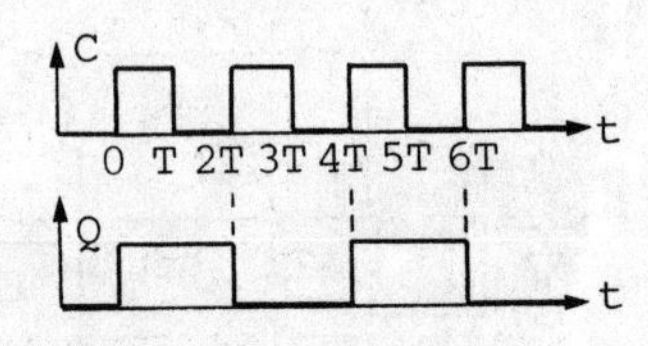

Waveforms when T=1

19.1.8 MASTER-SLAVE FLIP-FLOPS

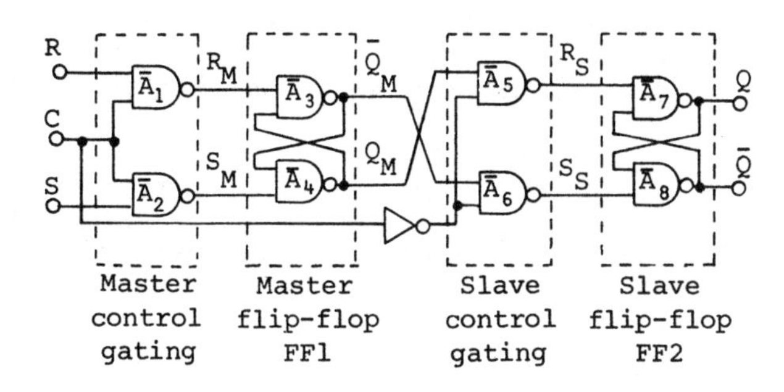

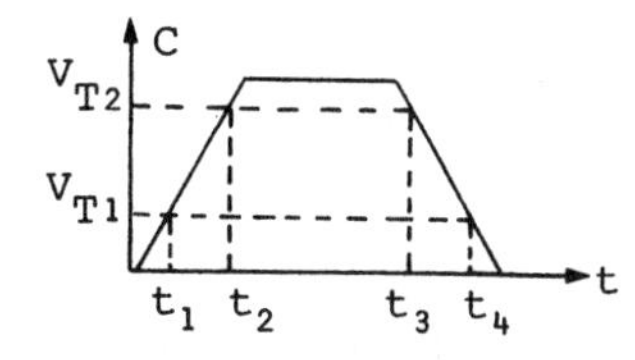

Summary of Operations in the M-S Flip-Flop	
Time	Operation
t_1	disable slave from master
t_2	enable master (slave remains disabled)
t_3	disable master (slave remains disabled)
t_4	enable slave

19.2 FLIP-FLOP TIMING

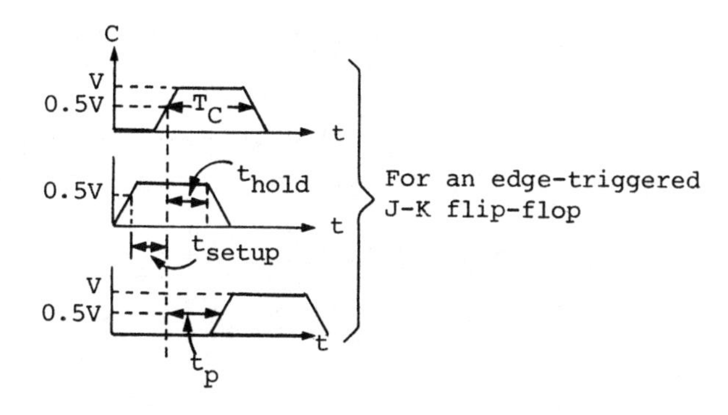

Hold, set-up and propagation time for a flip-flop:

The input data (J and K, R and S, or D) must be present and stable for some set-up time prior to the clock transition edge and for some hold time following the clock transition.

The clock transition edge is the rising edge of the clock pulse for a positive edge-triggered device and the falling edge of the clock pulse for a positive edge-triggered device.

Propagation time is measured from the 50% point on the clock transition edge to the 50% point on the output pulse edge.

19.3 COLLECTOR-COUPLED FLIP-FLOPS

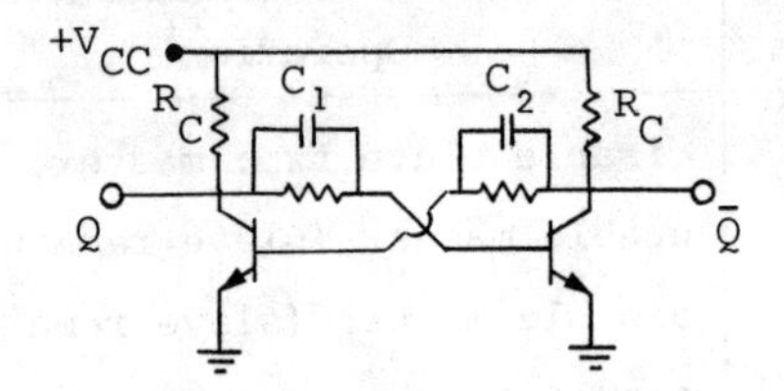

Collector-coupled ff

Test Conditions for a Flip-Flop

1) Do two stable states exist, in each of which at least one transistor is not active?

2) Is the incremental loop gain, with all transistors active, greater than 1?

With Q_1 off, $I_{B_2} = (V_{CC} - V_{BES_2})/(R_C + R_B)$. If Q_2 saturates, $I_{C_2} = (V_{CC} - V_{CES_2})/R_C$.

The condition for Q_2 to be saturated when Q_1 is off is:

$$I_{B_2} \approx (V_{CC} - V_{BES})/(R_C + R_B) \geq \frac{I_{C_2}}{\beta} = \frac{(V_{CC} - V_{CES})}{\beta \cdot R_C}$$

which is satisfied by any transistor with a minimum β,

$$\beta \geq \frac{V_{CC} - V_{CES}}{V_{CC} - V_{BES}} \cdot \frac{R_C + R_B}{R_C} \approx \frac{R_C + R_B}{R_C}$$

19.4 EMITTER-COUPLED FLIP-FLOPS

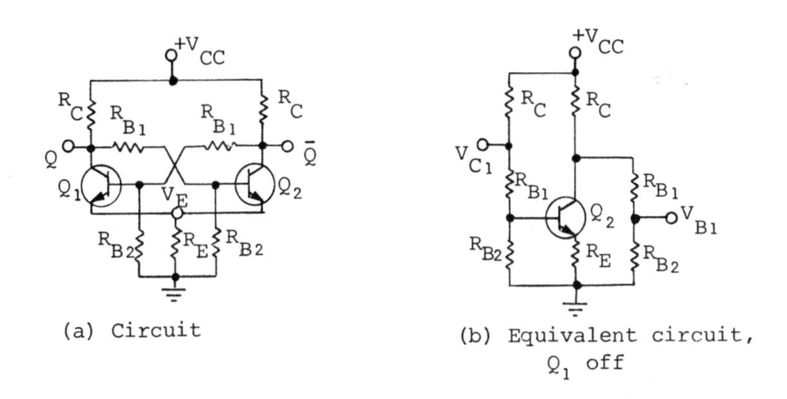

(a) Circuit

(b) Equivalent circuit, Q_1 off

With Q_1 off and Q_2 saturated, the emitter voltage is $V_E = (I_{B_2} + I_{C_2})R_E$. With Q_2 saturated, $V_{B_2} = V_E + V_{BES_2}$ and $V_{C_2} = V_E + V_{CES_2}$.

$$I_{C_2} = \left(\frac{V_{CC} - V_{C_2}}{R_C} + \frac{-V_{C_2}}{R_{B_1} + R_{B_2}} \right) \text{mA}$$

$$I_{B_2} = \frac{V_{CC} - V_{B_2}}{R_C + R_{B_1}} - \frac{V_{B_2}}{R_{B_2}}$$

Q_2 will saturate when Q_1 is off, if and only if $\beta \geq I_{C_2}/I_{B_2}$.

19.5 SWITCHING SPEED OF A FLIP-FLOP

Transition time: This is the time required for conduction to occur between two transistors. The transition time is reduced by using speed-up capacitors.

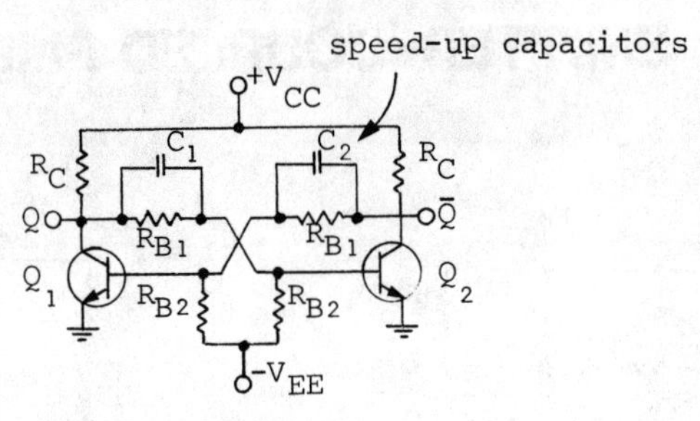

Fixed bias flip-flop

Without C_2 present, the input capacitance C_{i1} is at the base of Q_1 will limit the time constant with which v_{B_1} can rise, and thus the rate at which Q_1 can turn on to:

$$[R_C \,||\, r_{sat_2} + R_{B_1} + (R_{B_2} \,||\, R_{i_1})]C_{i_1}$$

A model for Q_1 is turning on: The optimum value of C_2 for which v_{B_1} rises in zero time to its final value with no overshoot or undershoot is

$$C_2 = \frac{R_{B_2} \cdot C_{i_1}}{R_{B_1}}$$

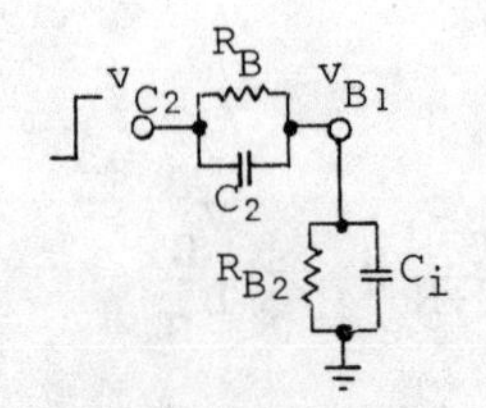

(a) Compensated attenuator

Equivalent circuit with Q_1 off and Q_2 saturated:

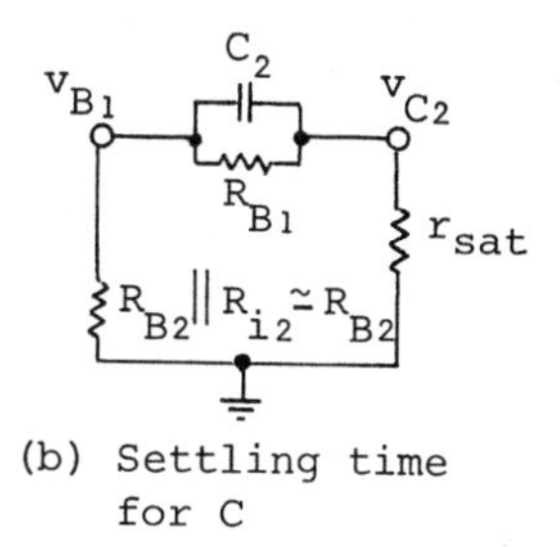

(b) Settling time for C

The voltage across C_2 changes at a rate

$$\tau_2 = [(R_{B_2} + r_{sat}) \| R_{B_1}] C_2,$$

$$\tau_2 \approx (R_{B_1} \| R_{B_2}) C_2 \text{ [with } r_{sat} \text{ very small]}$$

Equivalent circuit for determining the recharging rate for C_1:

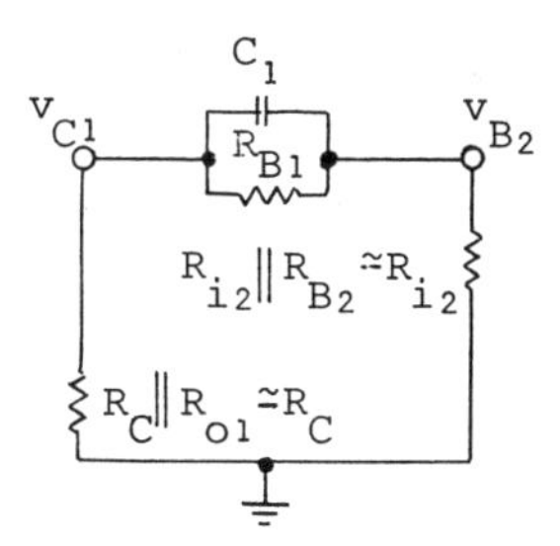

The value of the voltage on C_1 changes from its initial V_{C_2} value as soon as Q_1 has turned off. As Q_1 continues to be off, C_1 approaches its final steady-state value of

$$V_{C_2} = V_{C_1} - V_{B_2} \quad \text{at a rate} \quad \tau_1 = [(R_{i2} + R_c) \| R_{B_1}] \cdot C_1.$$

The settling time of the flip-flop: The time taken for C_1 and C_2 to recharge from their voltage levels ($V_{C_1} = V_{C_2} - V_{B_1}$ and $V_{C_2} = V_{C_1} - V_{B_2}$) to their new values ($V_{C_1} = V_{C_1} - V_{B_2}$ and $V_{C_2} = V_{C_2} - V_{B_1}$). (The values are interchanged.)

The resolution time is the sum of the transition time and the settling time. It is the total time the circuit requires to settle into its new steady state. It is a measure of the input trigger frequency that the circuit can resolve.

$$\text{The maximum frequency to which the flip-flop will respond} = \frac{1}{\text{The resolution time}}$$

19.6 REGENERATIVE CIRCUITS

Schmitt trigger:

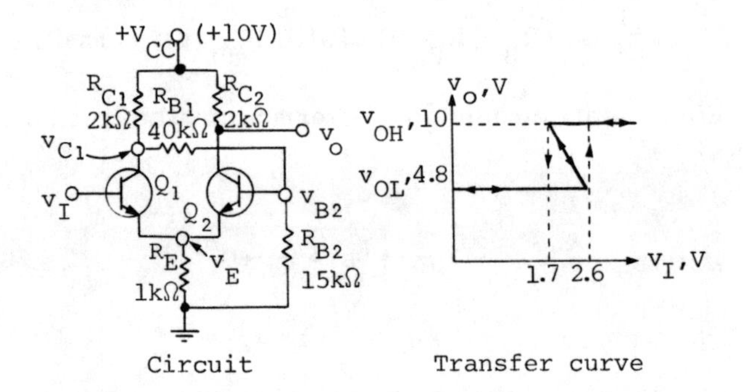

Circuit Transfer curve

Transfer curve:

The lower trace is the output response from $V_I = 0$ to 2.6V.

The maximum voltage for which Q_1 will remain off

$$V_{IL} = V_{B_2} = \frac{V_{CC} \cdot R_{B_2}}{R_{C_1} + R_{B_1} + R_{B_2}} = 2.6\text{v}.$$

While $V_I < \overline{V_{IL}}$, the output voltage is

$$V_o = V_{CC} - \frac{V_{B2} \cdot R_{C2}}{R_E} = V_{OL} \left(\text{where } I_{C2} \approx I_{E2} = \frac{V_E}{R_E} \approx \frac{V_{B2}}{R_E} \right)$$

$$= 4.8v.$$

When Q_1 is ON and Q_2 is OFF, the nodal equation at v_{C1} is

$$\frac{V_{CC} - v_{C1}}{R_{C1}} = i_{C1} + \frac{v_{C1}}{R_{B1} + R_{B2}}$$

The minimum $V_I = V_{IH}$ for which Q_1 shuts off occurs when $V_I = v_E = v_{B2}$. With Q_1 on and Q_2 off,

$$v_{B2} = \frac{v_{C1} \cdot R_{B2}}{R_{B1} + R_{B2}}$$

Hence,

$$v_{IH} = v_{B2} = \frac{V_{CC}}{R_{C1}} \left[\frac{1}{R_E} + \frac{1}{R_{B2}} + \frac{R_{B1} + R_{B2}}{R_{B2} \cdot R_{C1}} \right] = 1.7v$$

For $v_I > v_{IH} = 1.7$, $V_o = V_{CC}$.

For inputs in the range $\overline{v_{IL}} = 2.6V > V_I > 1.7 = v_{IH}$, the output will be either high or low, depending on its prior time history.

CHAPTER 20

WAVESHAPING AND WAVEFORM GENERATORS

20.1 COMMON WAVEFORMS

The step function: This function has an instantaneous change in level.

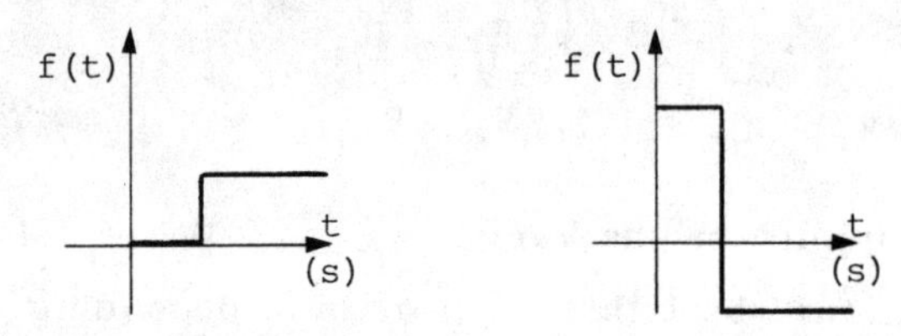

A ramp function:

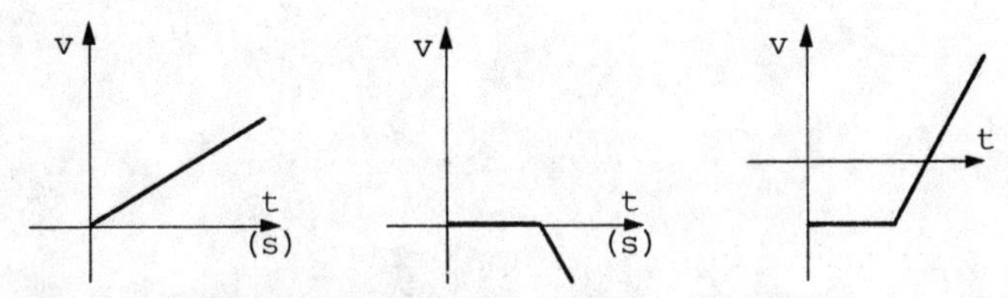

The exponential function:

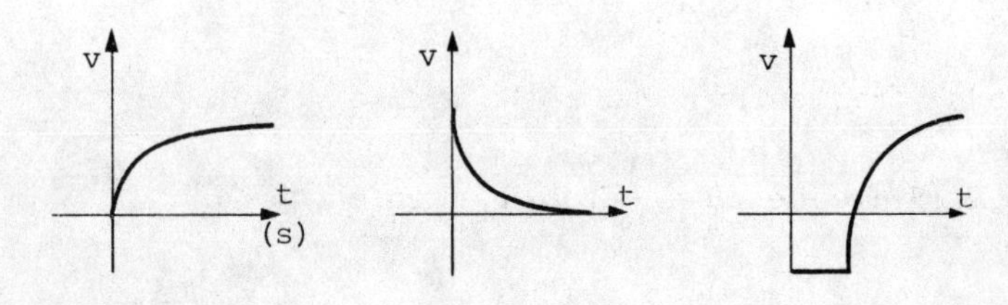

Pulse wave:

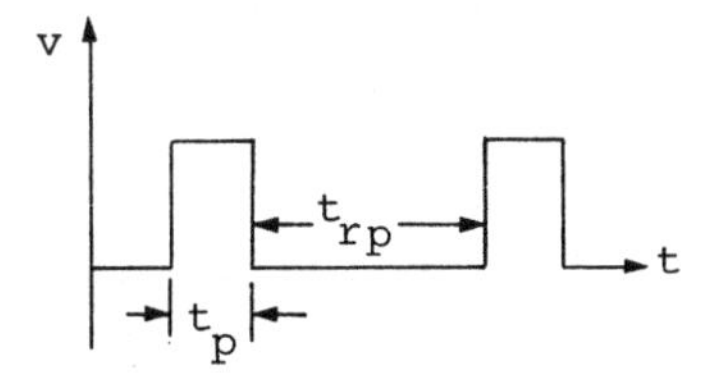

$$\text{Pulse repetition time (PRT)} = t_p + t_{rp} = T.$$

$$\text{Pulse repetition rate (PRR)} = \frac{1}{T}.$$

$$\% \text{ duty cycle} = \frac{t_p}{PRT} \times 100\%$$

20.2 LINEAR WAVESHAPING CIRCUITS

RC low-pass circuit:

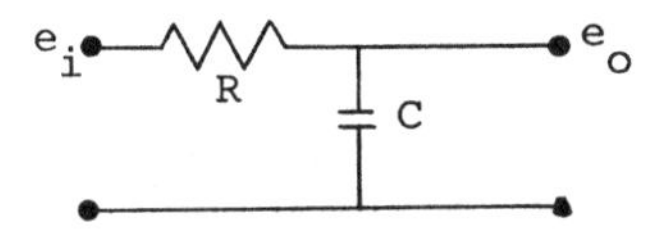

If the sinusoidal input e_i is applied, then

$$e_o = e_i \left(\frac{-jX_c}{R-jX_c} \right), \quad \text{where } X_c = \frac{1}{2\pi fc}$$

At low frequencies, $-jX_c >> R$, then $e_o \cong e_i$.

At high frequencies, $-jX_c << R$, then $e_o \cong 0$

Frequency response:

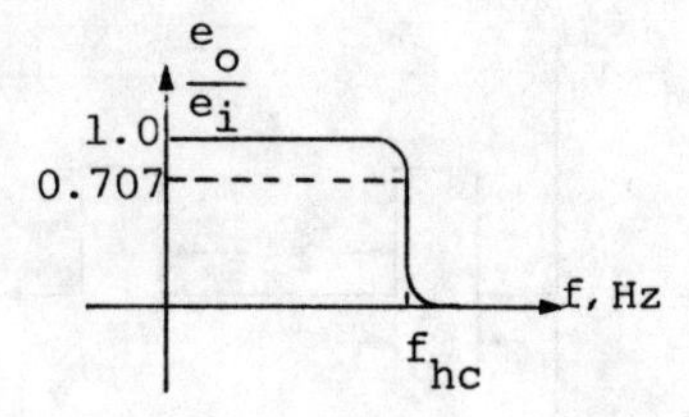

Cut-off frequency or half-power point frequency: The frequency at which the output (e_o) becomes 70.7 percent of the input e_i.

$$f_{nc} = \frac{1}{2\,\pi RC} = \frac{1}{2\,\pi\tau}, \quad \tau = \text{Time constant of the filter}$$

Step input to an RC low-pass:

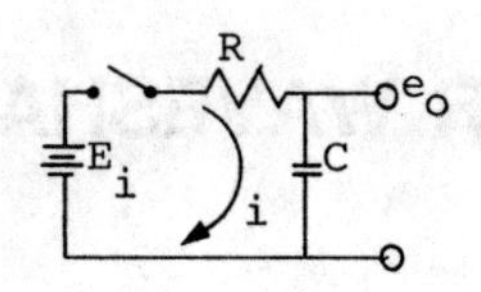

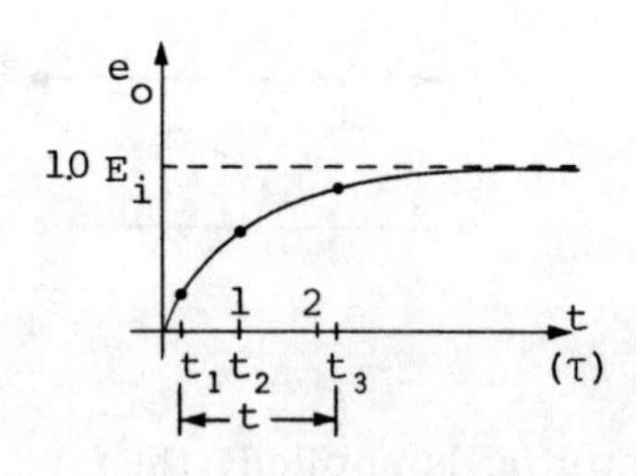

$$E_i = v_R + v_C = iR + V_C$$

$$= RC\frac{dv_c}{dt} + v_c$$

$$v_c = \begin{matrix}\text{The instantaneous}\\ \text{voltage across C}\end{matrix} = I_oR\left[1 - e^{-(t/\tau)}\right]$$

Curve characteristics:

$t_o = 0$ and $v_c = 0$ $\qquad$ $t_3 = 2.3\tau$, $v_c = 0.9E_i$

$t_1 = 0.1\tau$, $v_c = 0.1E_i$ $\qquad$ $t_4 = 5\tau$, $v_c \cong E_i$

$t_2 = \tau$, $v_c = 0.632E_i$

t_r = The rise time = $t_3 - t_1 = 2.2\tau$

Pulse wave input to an RC low-pass:

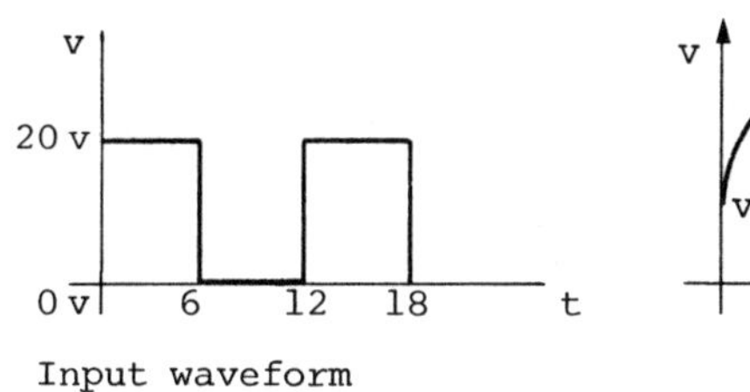

Input waveform

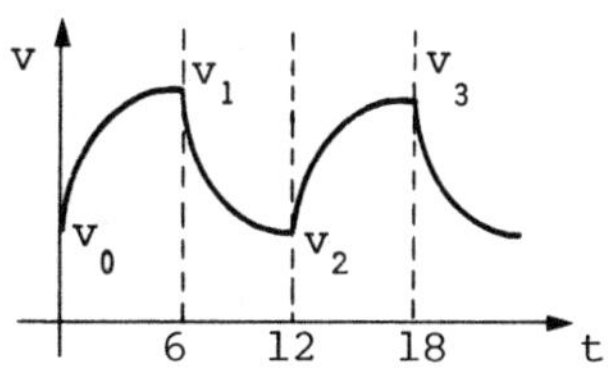

In steady state: $\quad v_o = v_2$ and $\quad v_1 = v_3$

$$v_1 = E_i - (E_i - v_o)e^{-(t/\tau)}$$

$$v_2 = E_i - (E_i - v_1)e^{-(t/\tau)}$$

Effects of the circuit τ on the output waveform:

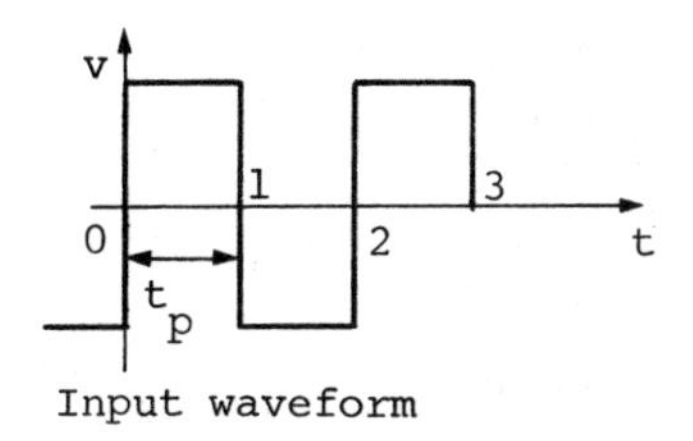

Input waveform

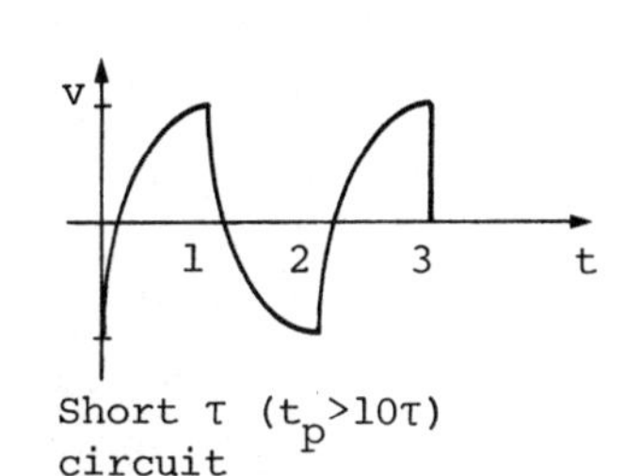

Short τ ($t_p > 10\tau$) circuit

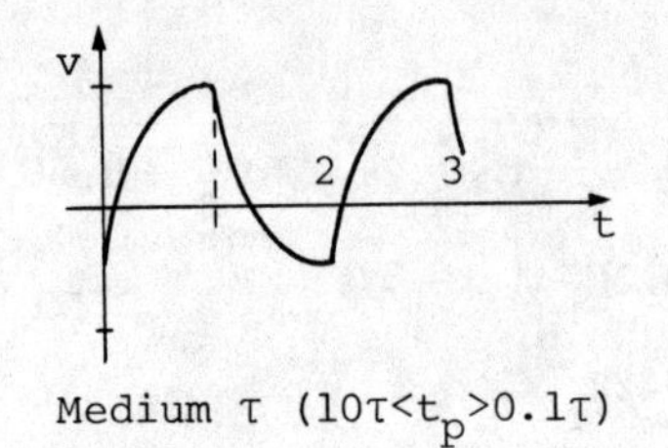

Medium τ ($10\tau < t_p > 0.1\tau$)

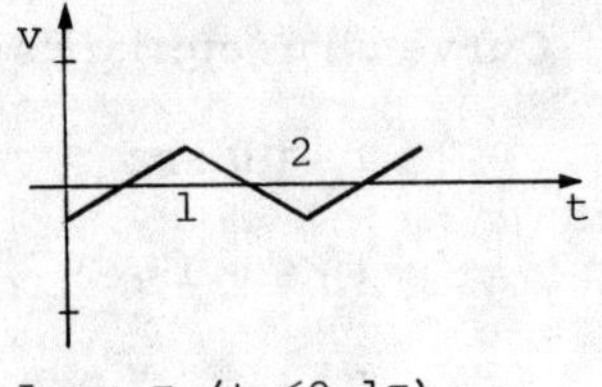

Long τ ($t_p < 0.1\tau$) circuit

RC high-pass circuit:

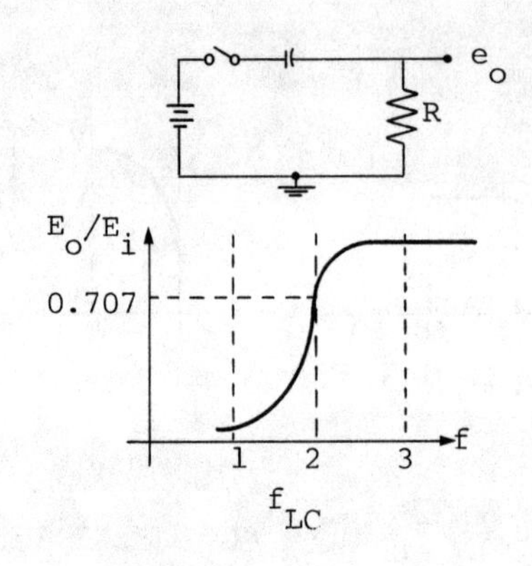

$$e_o = \left(\frac{R}{R - jx_c}\right) e_i, \quad x_c = 1/2\pi f_c$$

Lower-cutoff frequency $f_{\ell c} = 1/2\pi RC = \frac{1}{2\pi} \cdot \frac{1}{\tau}$

Pulse wave input:

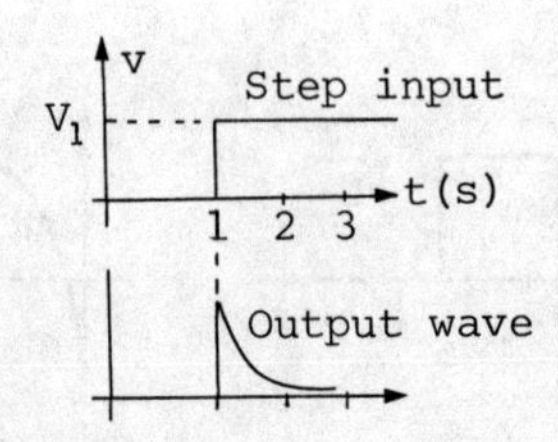

$$v_R = E_i - v_c = (E_i \pm V_{ci})e^{-(t/\tau)}$$

$$v_R = E_o = E_i\, e^{-t/\tau} \quad (\text{for } V_{ci} = 0)$$

Effects of τ on the output waveform:

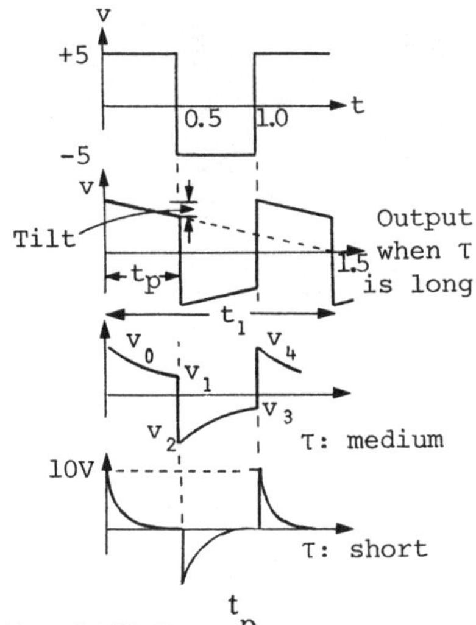

$$\text{Fractional tilt } F_t = \frac{t_p}{t_t}$$

(t_p = pulse width, t_t = Tilt time to reach 0 volt)

An R-C high pass circuit under short τ condition is called a differentiator.

RLC circuit:

Step input to a series RL circuit:

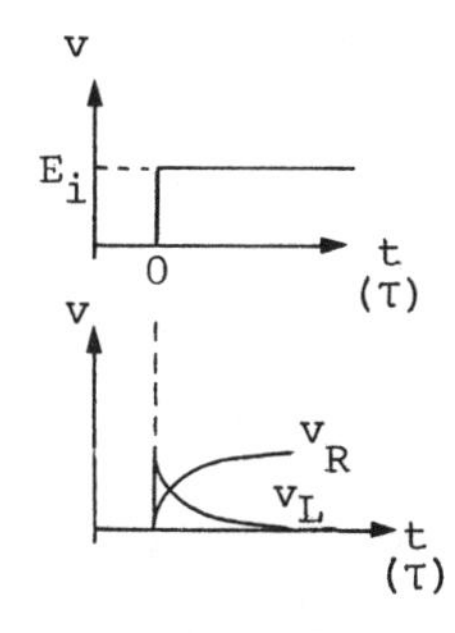

$$v_R = E_i \left(1 - e^{\frac{-tR}{L}}\right)$$

$$v_L = E_i \, e^{\frac{-tR}{L}}$$

Step input to a series RLC circuit:

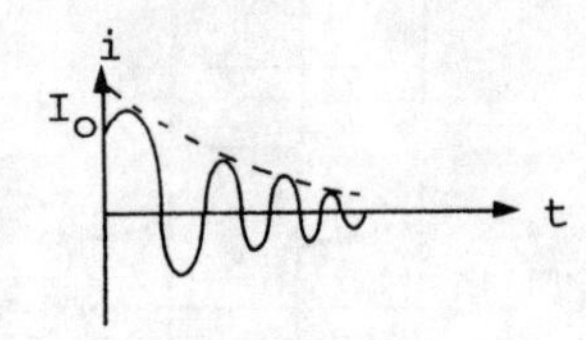

Current in a underclamped circuit

Attenuators:

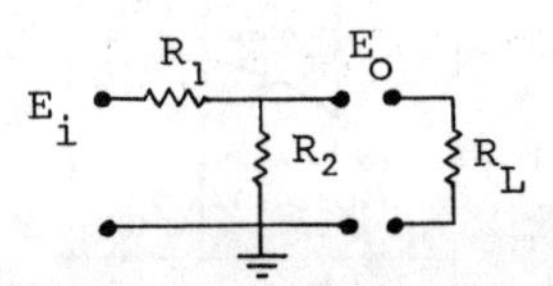

Simple attenuator

Attenuation factor: $\frac{E_i}{E_o} = \frac{R_2}{R_1+R_2}$

$$\frac{E_i}{E_o} \text{ (with } R_L\text{)} = \frac{R_2 \| R_L}{R_1 + (R_2 \| R_L)}$$

Uncompensated and compensated attenuator:

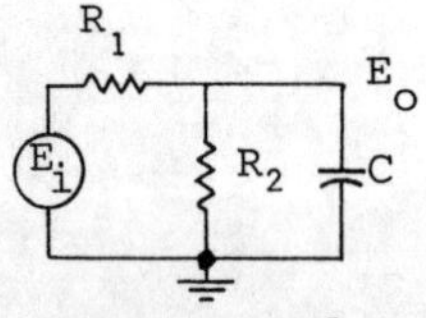

Uncompensated attenuator

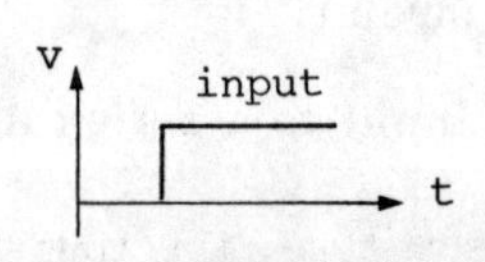

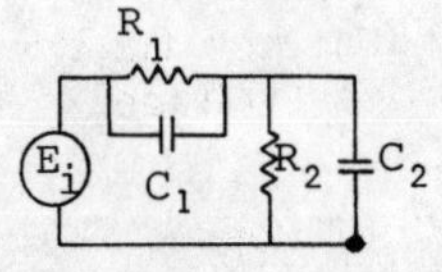

Compensated attenuator

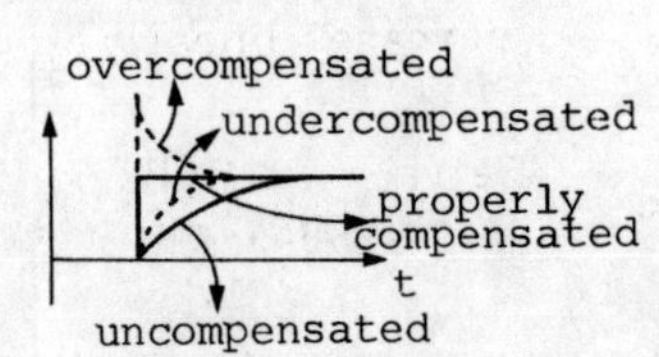

$$\frac{E_o}{E_i} = \text{Attenuation factor} = \frac{R_2 \| X_C}{R_1 + (R_2 \| X_C)}$$

$$\frac{E_o}{E_i} = \frac{Z_2}{Z_1 + Z_2}$$

$$Z_1 = R_1 \| X_{c1} \quad \text{and} \quad Z_2 = R_2 \| X_{c2}$$

For proper compensation:

$$R_1C_1 = R_2C_2$$

20.3 SWEEP GENERATORS

Voltage sweep principles:

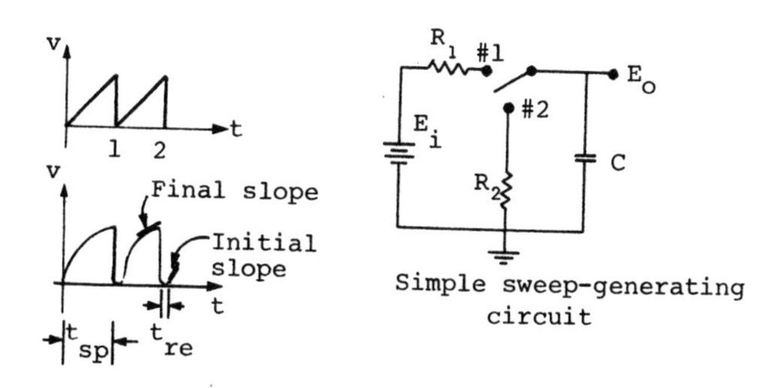

Simple sweep-generating circuit

t_{sp} = Sweep time, the time during which the function increases linearly.

t_{re} = Retrace or flyback time, the time the function takes to drop back down to the initial base voltage.

$$\%\ \text{slope error} = \frac{\text{Initial slope} - \text{final slope}}{\text{initial slope}} \times 100\%$$

Astable sweep circuit:

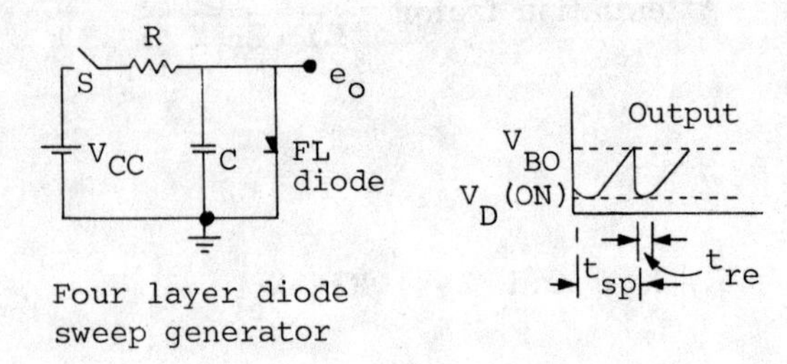

Four layer diode
sweep generator

$$V_{BO} = \text{(Breakdown voltage)}$$

$$= V_{CC} - [V_{CC} - V_D(ON)]e^{-(t_{sp}/\tau_c)}$$

$$t_{sp} = \tau_c \cdot \ln \frac{V_{CC} - V_D(ON)}{V_{CC} - V_{BO}} \text{ (s)}$$

$V_D(ON)$ = Voltage across FL diode when it is on

$\tau_C = RC$

Transistor sweep generator:

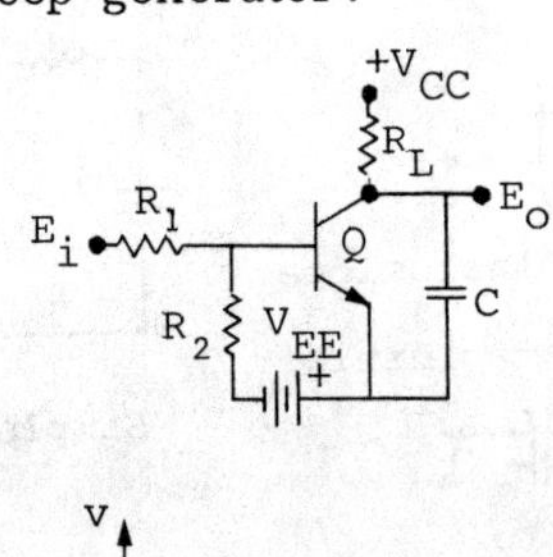

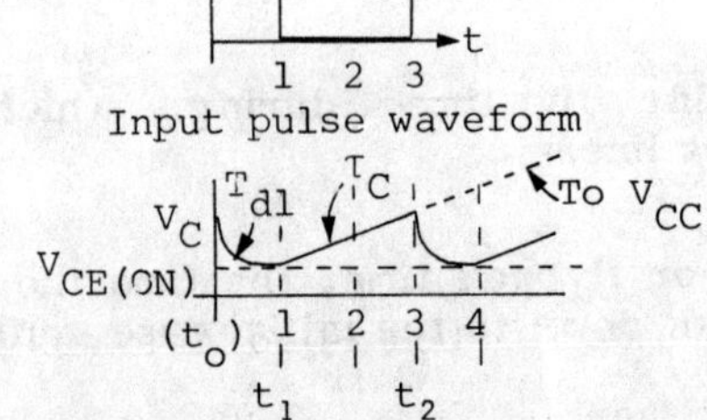

Input pulse waveform

V_C=voltage across capacitor just
before the transistor is ON

τ_C (charging time constant) = $R_L \cdot C$ seconds

τ_d (discharge time constant) = $R_T \cdot C$ seconds

R_T = Equivalent resistance of the ON transistor = $\dfrac{V_{CE}(ON)}{I_C(ON)}$ ohms

$$t_{sp} = t_2 - t_1 = \tau_c \cdot \ln\left[\frac{V_{CC} - V_{CE}(ON)}{V_{CC} - V_C}\right] \text{ seconds}$$

$$\text{Slope error} = \frac{V_C - V_{CE}(ON)}{V_{CC} - V_{CE}(ON)} \times 100$$

Miller sweep circuit:

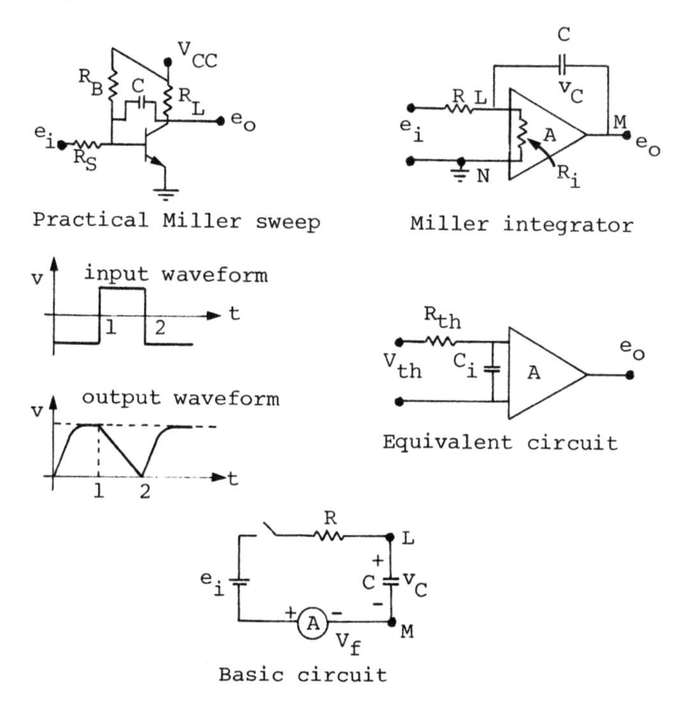

Basic circuit: V_f is a fictitious voltage and is made such that its terminal voltage is always equal to the voltage across C but opposite in polarity. In such a case, a constant current E_i/R will flow n the circuit.

$$V_c = E_i \frac{t}{\tau} \text{ volts}$$

Miller integrator: This simulates the basic circuit.

Equivalent circuit:

$$C_i = C(1+A)$$

$$V_{th} = E_i(R_i/R+R_i)$$

$$R_{th} = R_iR/R_i+R$$

$$V_{ci} = V_{th} - V_{th} \, e^{-(t/R_{th} C_i)}$$ ... Capacitor charge equation.

$$e_o = A \cdot e_i$$

$$= A \cdot V_{th} - A \cdot V_{th} \, e^{-(t/R_{th} C_i)}$$

The slope of the output for a high A:

$$A = \frac{e_o}{t} = \frac{E_i}{RC} = E_i \frac{t}{\tau_c}$$